FLORA OF TROPICAL EAST AFRICA

BORAGINACEAE

BERNARD VERDCOURT*

Annual or perennial herbs, often woody at the base, subshrubs, trees or woody climbers, usually characteristically scabrid or hispid. Leaves simple, alternate or less often opposite, petiolate or sessile, entire or variously toothed or crenate; stipules absent. Inflorescences terminal or axillary, bifurcate or in raceme-like or panicled groups, basically dichotomous cymes, the branches frequently scorpioid and spirally inrolled when young; bracts present or absent. Flowers ⚥ or unisexual (*Cordia*), sessile or pedicellate, mostly 5-merous, regular or somewhat irregular, occasionally heterostylous. Calyx tubular or campanulate, persistent and sometimes accrescent in fruit, with (2–)4–7 (or more) imbricate or rarely valvate or open lobes. Corolla tubular, campanulate, funnel-shaped or rotate with tube very reduced; lobes 3–16, imbricate or contorted (*Myosotis*); throat often with scales, thickenings, folds or crests. Stamens the same number and alternating with the corolla-lobes, exserted or included, epipetalous; anthers 2-thecous. Nectariferous disc usually present. Ovary superior, entire or deeply 4-lobed, 2-locular with 2 ovules in each or with 4, 1-ovuled locules due to the development of false partitions. Ovules axile, anatropous, erect, horizontal, basal or rarely pendulous. Style usually 1, 2 in one genus or sometimes 4 in one species of *Cordia*, terminal or gynobasic, entire or 2–4-fid or twice 2-fid; stigmas clavate, capitate or peltate or linear to narrowly foliaceous. Fruit drupaceous with 1, 4-celled endocarp or separating in 2 bilocular or 4 unilocular 1-seeded pyrenes or with 4 quite distinct nutlets. Seed with or without endosperm, erect, oblique or horizontal; testa membranous; embryo straight or curved; cotyledons flat or plicate.

A considerable family occurring in both temperate regions and tropics of Old and New Worlds with perhaps 2000 species (about three-quarters herbs and subshrubs, the rest trees and shrubs) in about 100 genera; often now divided** into two families, Boraginaceae proper with mainly herbs and Ehretiaceae nom. conserv. (*Cordiaceae, Sebestenaceae*) with mainly trees and shrubs but some quite herbaceous genera. To me the family is unquestionably a natural unit. The *Heliotropioideae* and *Wellstedioideae* have also been regarded as separate families Heliotropiaceae and Wellstediaceae.

The family does of course contain many plants widely grown in gardens in temperate regions. Some of them are dealt with under their respective genera. Of those belonging to genera not occurring wild or naturalised in East Africa the following may be mentioned — *Borago officinalis* L., 'borage', has been grown at Njombe, originating from a packet of 'mixed herbs' seeds (Njombe, 26 May 1967, *S.A. Robertson* 661!); it has also been seen from the W. Usambaras but whether naturalised or cultivated is not certain (near Lushoto [Wilhelmstal], Jaegertal, 25 May 1914, *Peter* 52341!).

H.H. Hilger is carrying out detailed studies of the development of flowers and fruits; for an account of the tribes *Cynoglosseae* and *Eritrichieae* see E.J. 105: 323–378 (1985).

1. Trees, shrubs or sometimes lianes with globose to ovoid or
   ellipsoid drupe-like fruits; stigmas 1–4, terminal
   (*Ehretioideae* pro parte and *Cordioideae*)   2

---

* Great use has been made of the series of papers by I.M. Johnston in the Journ. Arn. Arb., the account of the family for Flore du Congo, Rwanda et Burundi by A. Taton (1971) and that by H. Heine for Flore de la Nouvelle Calédonie 7 (1976); also accounts in several recent Middle East floras. I have not hesitated to use some descriptions almost verbatim or with modifications needed by the examination of more extensive material. My thanks are due for this plagiarism carried out without previous permission.

** See Cronquist, Int. Syst.: 917 (1981) who keeps them together and Hutchinson, Fam. Fl. Pl., ed. 3: 485 (1973) etc., who keeps them apart.

Herbs or subshrubs, the fruit mostly of 4 nutlets but, if
    trees, shrubs or lianes with drupe-like fruits then style
    with stigmatose ring below bifid apex (see couplet 7) 4
2. Style twice bifid with 4 stigmas; calyx usually accrescent and
    cupular in fruit . . . .   **1. Cordia**
    Style bifid with 2 stigmas or in a few species undivided and
    with ± bifid stigma; calyx not accrescent 3
3. Calyx-lobes imbricate; outer surface of pyrenes rugulose,
    irregularly reticulate or almost smooth . . . . .   **2. Ehretia**
    Calyx-lobes valvate; outer surface of pyrenes with close
    ridges bearing overlapping wings . . . .   **3. Bourreria**
4. Style(s) terminal on the ovary 5
    Style inserted between the very distinct ovary-lobes
    (*Boraginoideae*) 8
5. Styles 2, ± joined at base; prostrate herb with crenulate or
    lobulate leaves (*Ehretioideae* pro parte) . . . . .   **4. Coldenia**
    Style 1 with stigmatic ring below the sometimes bifid apex
    or stigma sometimes ± terminal (*Heliotropioideae*) 6
6. Fruit consisting of 1–4 separate dry nutlets . . . . .   **7. Heliotropium**
    Fruit drupe-like, fleshy or ± dry, globose with 4 1-seeded
    pyrenes often cohering in pairs 7
7. Silvery-grey hairy shrub or small tree of the shore; fruit ±
    dry when mature, with spongy mesocarp . . . .   **5. Argusia**
    Green climbing shrub or herb of rain-forest; fruit
    fleshy . . . .   **6. Tournefortia**
8. Flowers irregular 9
    Flowers regular 10
9. Stamens long-exserted; annual casual (*Lithospermeae*)   **10. Echium**
    Stamens included; indigenous perennial subshrub of ±
    arid areas (?*Eritrichieae*) . . . .   **13. Echiochilon**
10. Anthers included or visible at the throat or very shortly
    exserted, not or only with short appendages; calyx-
    lobes small, not cordate or hastate at base; nutlets
    smooth or rough 11
    Anthers very distinctly exserted, with distinct appendages,
    often connivent, or if hidden in tube (in one
    Trichodesma) then calyx-lobes large, 0.9–1.3 cm. long,
    cordate to hastate at the base; nutlets rough, margined,
    tuberculate etc. 15
11. *Cynoglossum*-like plant; corolla white with 10 distinct
    nectaries at base; bosses at corolla-throat and stamens
    more distinctly exserted; style of fertilised flower up to
    7 mm. long; nutlets unknown in E. African species
    (T4) . . . .   **16. Afrotysonia**
    Corolla white or blue without such distinct nectaries;
    bosses and stamens not or less exserted; style usually
    under 3 mm. long 12
12. Nutlets smooth and shining 13
    Nutlets verrucose or partly to entirely glochidiate 14
13. Flowers blue; corolla-lobes contorted; anthers placed at
    top of usually short tube; connective with minute apical
    appendage (*Eritrichieae*) . . . .   **12. Myosotis**
    Flowers white; corolla-lobes imbricate; anthers placed at
    middle of more distinct tube; connective with ±
    imperceptible appendage (*Lithospermeae*) . . . .   **8. Lithospermum**
14. Nutlets ± globose or compressed, partly or completely
    covered with glochidia; corolla-throat with conspicuous
    bosses ("scales"); annual or perennial (widespread
    indigenous plants) . . . .   **15. Cynoglossum**
    Nutlets pyriform, beaked, rugulose-tuberculate; corolla-
    throat without bosses but with 5 longitudinal hairy
    bands (introduced weed) . . . .   **9. Buglossoides**

15. Filaments with marked narrow appendages; annual very rough herb with rotate corolla and nutlets with thickened collar at base (cultivated) (*Boragineae*)   **Borago**

Filaments not appendaged but with inflated gibbosities; anthers with very distinct often twisted thin apical appendages; corolla subrotate or funnel-shaped; nutlets variously margined, roughened or tuberculate 16

16. Anther-appendages mostly twisted save in one species; filaments flattened and glabrous; calyx accrescent (*Cynoglosseae**) . . . .   **14. Trichodesma**

Anther-appendages not twisted; filaments with enlarged gibbosities and hairy at base; calyx not accrescent (*Lithospermeae*) . . . .   **11. Cystostemon**

1. CORDIA

L., Sp. Pl.: 190 (1753) & Gen. Pl., ed. 5: 87 (1754); Warfa, Acta Univ. Upsal. 174: 1–78 (1988)

Trees or shrubs, less often climbers or scramblers. Leaves alternate or in a few (but sometimes very common species) subopposite, petiolate, simple, often large, entire to crenate-dentate. Flowers mostly white, yellow or orange, ♂, polygamous or unisexual (plant dioecious), subsessile or pedicellate, borne in terminal or axillary dichotomous corymbs, panicles or subglobose clusters of cymes, the branches scorpioid, without bracts. Male flowers with 4–8 stamens, the filaments often hairy at the base; ovary rudimentary but style absent. Female with anthers sterile, otherwise similar to ♂ flowers. Calyx tubular or campanulate, smooth or with marked ribs, 2–5(–more)-lobed, persistent and accrescent in fruit. Corolla funnel-shaped or salver-shaped, mostly 5- but sometimes 3–8-lobed; tube short or long, cylindric or widened; lobes erect, spreading or reflexed, imbricate or subcontorted in bud; stamens exserted or included, the filaments glabrous or pubescent at the base. Ovary 4-locular with 1 erect ovule in each locule. Style terminal, twice bifid (or abnormally twice trifid), the ultimate stigmatic parts of the 4 branches linear to subfoliaceous or terminated by 1 capitate or peltate stigma, rarely with 4 separate styles. Fruits ovoid, globose or ellipsoid, included in or sitting in the persistent accrescent cupuliform calyx; endocarp bony with up to 4 locules but only 1–2 fertile. Seeds without endosperm.

About 250–300 species in the tropics of both hemispheres but mainly American. Very many of the species are of great ecological significance and some are important browse shrubs in dry country. It is a heterogeneous genus and it has been suggested it should be split; there is certainly tremendous diversity, particularly in S. America. Linnaeus included 3 species in the Species Plantarum, *C. myxa, C. sebestena* and *C. glabra*. *C. sebestena* has been chosen as the type of the genus by many workers (Hitchcock & Green, (Int. Bot. Congress Cambridge 1930, Nom. Prop: 133 (1929)), Phillips, (Gen. S. Afr. Fl. Pl., ed. 2: 627 (1951)) and *C. glabra* by Britton and Wilson, (Bot. Porto Rico & Virgin Is. (2): 125 (1925)) who split the genus. Friesen (in Bull. Soc. Bot. Genève, sér. 2, 24: 117–201 (1933)) discusses the matter and considered that the description of *Cordia*, derived from material of both Plumier and Dillenius, did not agree with the genus as known and excluded it from the family; he recognised 12 genera in what he called the Sebestenaceae including several new ones. It is true Plumier (Nov. Gen. Amer.: 13, t. 14 (1703), shows a bifid stigma and a bilocular ovary and Linnaeus's description repeats this information. I.M. Johnston (Journ. Arn. Arb. 16: 4, 5 (1935)), quite rightly concluded that the drawings were faulty; the original Plumier drawing in Paris undoubtedly depicts *C. sebestena*.** Unfortunately only a few species belong to section *Cordia* if this is accepted as genotype. The acceptance of Britton and Wilson's selection of *Cordia glabra* L. appears fraught with difficulty. Linnaeus, Sp. Pl.: 191 (1753) gives a description under this which would indicate a specimen but none has been found; a queried reference to Plumier, Gen. 13 is clearly of no consequence; later in Sp. Pl., ed. 2: 275 (1762) Linnaeus included *C. glabra* in the synonymy of *C. collococca* L. thus rendering that name illegitimate or so one might suppose. I.M. Johnson (Journ. Arn. Arb. 21: 345 (1940)), however, expressed the opinion that this placing in synonymy was a clerical error and that *C. glabra* is actually a synonym of *Bourreria succulenta* Jacq. (= *Ehretia bourreria* L.) since the description of the latter exactly repeated that of *Cordia glabra*. Since the name *Bourreria glabra* already existed no action was

---

* Riedl (Notes Roy. Bot. Gard. Edin. 40: 2 (1982)) suggests "it probably should be placed in the separate tribe *Trichodesmeae*"

** It must be pointed out that under the existing rules the type of *Cordia* is the type of the species chosen to typify it and for none of the three included by Linnaeus is any Plumier reference cited.

necessary. In the absence of a type it might seem best not to tamper with Linnaeus' own actions but since the doubt has been raised *Cordia glabra* is hardly a suitable lectotype in the circumstances. If the genus were split it would be wise to retypify it with *C. myxa* as type in order to avoid as many changes in name as possible. Nowicke & Ridgway (Amer. Journ. Bot. 60: 584–591 (1973)), demonstrated that there were at least 3 well-marked pollen types in the genus as at present constituted. With the aid of Miss C. Tingay I extended this to most E. African species which confirmed two groups for the Old World. The evidence for splitting the genus seems quite substantial, particularly as *Cordia* species proper are large-flowered species with the fruit completely enveloped in the calyx. Very recently, however, Taroda & Gibbs (Rev. Brasil. Bot. 9: 31–42 (1986)) have re-examined the problem after study of the S. American species and decided that although the division into 3 pollen types is clear-cut, correlations with floral characters are not. There are many genera where pollen variation is surprisingly wide but is never suggested as sufficient evidence for generic segregation, e.g. *Saxifraga*. More recently still Gaviria (Mitt. Bot. Staats., München 23: 1–273 (1987)) has monographed the Venezuelan species and also retained the genus in its traditional sense. I have therefore decided to follow the new infrageneric classification put forward by Taroda & Gibbs who discuss the matter in detail. Two subgenera occur in East Africa.

Subgen. **Cordia**.* Calyx smooth, markedly accrescent, totally enclosing and adhering to the fruit. Corolla large, orange-red. Pollen grains with exine striato-reticulate. Sp. 1 and cultivated *C. sebestena***

Subgen. **Myxa** *Taroda*. Calyx smooth or ribbed, sometimes slightly accrescent, partially enclosing the fruit. Corolla smaller or large, mostly whitish, marcescent and enclosing fruit in one section. Pollen grains with spinulose exine.

Sect. **Myxa** (*Endl.*) *DC*. Calyx ± smooth save in one species. Corolla fairly small, not marcescent nor enclosing fruit. Spp. 2–28.

C. africana has a very ribbed calyx similar to many S. American species of sect. *Gerascanthus* (Browne) Don but pollen data (C. Tingay & I.K. Ferguson, personal communication) does not support treating it as a separate Old World section.

A number of species are cultivated in East Africa. *C. sebestena* L. (U.O.P.Z.: 210, fig. (1949); Dale, Introd. Trees Uganda: 25 (1953); Jex Blake, Gard. E.Afr., ed. 4: 109 (1957); Gaviria in Mitt. Bot. Staats., München 23: 33 (1987)). Native of the W. Indies. Small tree 3–7.5(–10) m. tall with terminal spike-like cymes of large bright red or orange flowers, ovate, elliptic or rounded thick rough leaves, 7–30 cm. long, 4–15.5 cm. wide; calyx ± 2 cm. long, lobed, with brown pubescence and white setae; corolla funnel-shaped, (2–)4–5 cm. long, 3.8 cm. wide with 6 rounded lobes; fruit ovoid, pointed, 2–4 cm. long. This is a commonly cultivated tree and material has been seen or reported from Uganda, Makerere (*Lye* 6055); Kenya, Nairobi; Tanzania, Kilosa (*Sangiwa* 65), Morogoro (*Wigg* 1069), Dar es Salaam (*Harris* 4444), Kilwa, Masoko (*Ruffo* 361) and Zanzibar. *C. dodecandra* DC. Native of Central America. Shrub or small tree with elliptic leaves, 18 cm. long, 12 cm. wide, very rough above, densely white-pubescent beneath and red flowers 4 cm. long, the limb white-pubescent inside and out. Tanzania, Mombo (*Borota* 79!). *C. lutea* Lam., an ornamental shrub from Peru ± 1 m. tall with yellow flowers is rather similar to *C. africana* in having a very strongly ribbed calyx but the leaves are rough above; there are no stellate hairs. Three specimens have been seen from Dar es Salaam, Ikulu (*Ruffo* 1056!) & State House, 22 Sept. 1972, *Ruffo & Pallangyo* 2743! & 24 Nov. 1972, *Ruffo* 564! *C. alliodora* (*Ruiz & Pavon*) *Oken*** (T.T.C.L.: 75 (1949)). Native of Central America. Shrub or small tree 1.8–20 m. tall with elliptic-obovate to elliptic-oblong acuminate leaves 16 cm. long, 8 cm. wide which are alternate or subopposite, stellate-pubescent on both sides when young, becoming glabrous above; calyx 5 mm. long, conspicuously ribbed; flowers in extensive stellate-tomentose panicles; corolla white; tube 5–6 mm. long, lobes oblong, 7 mm. long, 3 mm. wide. Tanzania, Amani (*Greenway* 2866! & 7021!; *Ngonyani* 17!) and Lushoto Arboretum (*Kyessi* 29! & 30!).

1. Leaves opposite or subopposite, rarely some alternate**** 2
   Leaves essentially alternate   3

---

* See note on p. 3.

** Taroda & Gibbs are incorrect in stating that only one wild Old World species belongs to this subgenus; *C. suckertii* Chiov. (Somalia S.) undoubtedly does.

*** Authority usually given as Chamisso who though earlier did not actually make the combination.

**** Omitted from the key are species 9 (*C. fischeri*), species 12 (*C. sp. B*, near *C. crenata*), species 26 (*C. sp. E*) and species 27 (*C. sp. F*), of which no material has been seen or is entirely inadequate; *C. sp. E* is mentioned on p. 7 (footnote).

17. Littoral or cultivated shrub with large white to orange or
    red flowers with tube 2–3 cm. long; leaves 4–34 × 3–17
    cm.   . . . .   return to couplet 7
    Shrubs or trees, if littoral then with much smaller white or
    cream flowers   18
18. Trunk angular and twisted with characteristic smooth
    peeling grey-white bark; fruits small, 8–10 × 6 mm.;
    leaves glabrous or with dense fine tubercle-bases only
    (in Flora area)   . . . .   3. *C. goetzei*
    Trunk and bark different or not known or, if similar, leaves
    pubescent and/or rough above   19
19. Foliage, etc. ± entirely glabrous   . . . .   3. *C. goetzei*
    Foliage not entirely glabrous   20
20. Leaves almost always very rough above like sandpaper;
    corolla 0.8–1.4 cm. long; fruit 0.8–2 × 0.6–1.2 cm.   21
    Leaves never so rough   22
21. Cymes several-flowered   . . . .   8. *C. monoica*
    Fruits sessile and solitary in only known ± complete
    specimen   . . . .   11. *C. sp. A*
22. Tree to 18 m. or shrub to 5 m.; leaves prominently 3-nerved
    from the base, at length glabrous above, velvety
    beneath (K7, one sterile specimen; T3, East Usambaras,
    known from one fruiting specimen)   . . . . . .   23. *C. torrei*
    Without above characteristics   23
23. Shrub 1–2 m. tall with many short leafy and flowering
    shoots or inflorescences with leaves still in a very
    juvenile state; young shoots, inflorescence-axes,
    undersides of leaves etc. with dense closely adpressed
    buff indumentum; leaves at length very discolorous;
    calyx-tube 6 mm. long (T8)   . . . .   24. *C. trichocladophylla*
    Mostly flowering when leaves developed but if precocious
    then calyx-tube about 2 mm. long and indumentum
    different   24
24. Shrubs with flowers appearing while only undeveloped
    leaves are present at apices of shoots, otherwise
    apparently leafless; ♂ calyx campanulate; tube 2.2 mm.
    long, lobes broadly triangular 1.3 mm. long; corolla-
    tube 2.2 mm. long; lobes oblong to oblong-lanceolate,
    3.5–4 × 2 mm. (virtually unknown *Brachystegia* woodland
    species from T8)   . . . .   19. *C. sp. D*
    Developed leaves and flowers occurring together   25
25. Leaves mostly large, 3.5–33 × 2–26 cm.; fruits (1.2–)2–4.4
    cm. long but not known in *C. sp.* 4 or 18   26
    Leaves mostly smaller; fruits 0.9–1.2 cm. long   30
26. Young leaves and calyx with well-spaced adpressed short
    white hairs; calyx-tube obconic, ± 5 mm. long; corolla-
    tube 5 mm. long (T3, East Usambaras, known from one
    specimen only)   . . . .   4. *C. peteri*
    Not as above   27
27. Inflorescence-branches and stems covered with dense
    long pale brownish hairs; inflorescences lax, 9 × 8 cm.;
    leaves rounded ovate, ± 12.5 × 10 cm., glabrous save for
    a few sparse long hairs (virtually unknown species
    from T8)   . . . .   18. *C. sp. C*
    Not as above; leaves often velvety with tertiary venation
    raised and closely parallel but occasionally glabrous in
    cultivated species   28
28. Calyx 1.2–2 cm. long in ♀ flowers; leaves with toothed
    margins, velvety (Tanzania)   . . . .   5. *C. fissistyla*
    Calyx much shorter   29

29. Stigma-arms flat and subfoliaceous; leaves 3–18 × 3–20 cm., glabrous to densely velvety beneath, sometimes crenate; fruit (1.2–)2–3.5 cm. long (introduced) . . . . 2. *C. myxa*

Stigma-arms linear; leaves 3.5–33 × 2–26 cm., with venation raised beneath, the tertiary veins closely parallel and ± velvety; fruits 2–4.4 cm. long (mostly tall native timber tree, Uganda, W. Kenya, W. Tanzania) . . . . . 6. *C. millenii*

30. Petioles long, often almost ¾ the length of the blade; leaf-blades triangular-ovate, glabrescent; inflorescence-axes floccose and with longer hairs; calyx-tube ± glabrous . . . . 14. *C. longipetiolata*

Petioles much shorter or if up to ½ length of blade then calyx densely pubescent 31

31. Leaves mostly obovate; fruit very sharply tapering and pointed, up to 1.7 cm. long; mostly littoral or at least coastal plant . . . . 17. *C. somaliensis*

Leaves not so distinctly obovate; fruit and habit not as above (save in 20. *C. guineensis*) 32

32. Leaves rather thick, mostly distinctly 3-nerved from the base; calyx rather thick and slightly ribbed; fruit tapering-acuminate (**K7, T6, Z**) . . . . 20. *C. guineensis**

Leaves thinner, less distinctly 3-nerved at base, often crenate-dentate; calyx thin, not ribbed; fruit less acuminate . . . . 13. *C. crenata*

1. **C. subcordata** *Lam.*, Tab. Encycl. 1: 421 (1792); Poir. in Encycl. Méth. Bot. 7: 41 (1806); A.DC. in DC., Prodr. 9: 477 (1845); Vatke in Linnaea 43: 314 (1882); C.B.Cl. in Fl. Brit. Ind. 4: 140 (1883); Bak. & Wright in F.T.A. 4(2): 10 (1905); U.O.P.Z.: 212 (1949); T.T.C.L.: 76 (1949); I.M. Johnston in Journ. Arn. Arb. 32: 3 (1951); K.T.S.: 71 (1961); E.P.A.: 769 (1961); Heine, Fl. Nouv.-Caléd. 7: 104, t. 23 (1976); Fosberg & Renvoize, Fl. Aldabra: 194, fig. 31/1–3 (1980); Nowicke, Miller & Bittner in Journ. Palynol. 23–24: 59–64 (1988); Warfa, Acta Univ. Upsal. 174: 51, fig. 2/C, 4/B (1988); Martins in F.Z. 7(4): 62 (1990). Lectotype chosen by Fosberg (annotation on sheet): Seychelles, Praslin, *Commerson* (P-JU 6479, lecto.!)

Shrub or tree, 1.5–15 m. tall with spreading branches; trunk thick, up to 60 cm. diameter; stems pale grey, ± ridged; bark brown, ± rough. Leaf-blades broadly ovate to almost round, 4–34 cm. long, 3–17 cm. wide, rounded to acute or subacute or shortly acuminate at the apex, rounded at the base or sometimes ± subcordate or obtuse or quite tapering, slightly succulent, very shortly pubescent above when young but later ± glabrous above but often with cystolith spots, velvety to ± tomentose along edges of costa and nervation beneath or ± glabrous; petiole 1.5–11 cm. long. Inflorescence axillary or pseudo-terminal 6–20-flowered cymes, 10–14 cm. long, pubescent when young; peduncles 0.5–2 cm. long; pedicels 2–5(–10) mm. long; flowers heterostylous. Calyx brownish, cylindric-funnel-shaped, 1.1–2 cm. long, 3–5-valvate or -lobed, the lobes 2–5 mm. long, finally accrescent and completely enveloping the fruit, glabrous or slightly pubescent outside. Corolla white or usually apricot, orange or red, funnel-shaped; tube 2–3 cm. long, 3 mm. diameter at base, 1–2 cm. wide at the throat; lobes 5–7, rounded to obovate-spathulate, 1.5–2.5 cm. long and wide, spreading, usually crinkly-edged. Filaments inserted just below the middle of the corolla-tube, the anthers level with the throat. Style** 2–2.5 cm. long (short) or ± 3 cm. long (long); primary branches 2–7 mm. long, the 4 stigmatic lobes 0.5–2 mm. long, flattened. Fruit yellow to red, becoming blackish, ovoid, obovoid or subglobose, 1.5–4 cm. long, 2–3 cm. wide, sometimes ± beaked, the enveloping calyx becoming hard; endocarp very hard, angular and furrowed, 4-locular but only 2 seeds developing. Fig. 1/1,2, p. 8.

Kenya. Kwale District: coast E. of Waa, 4 July 1975, *Gillett* 20837!; Kilifi District: 32 km. N. of Mombasa, Vipingo, 16 Dec. 1953, *Verdcourt* 1070!; Lamu District: Kiunga, 29 July 1961, *Gillespie* 87!

* See also sp. 26 *C. sp. E* which is similar to *C. guineensis* but has a different facies.
** The 23–70 cm. given by Heine must be wrong, I think, even if mm.

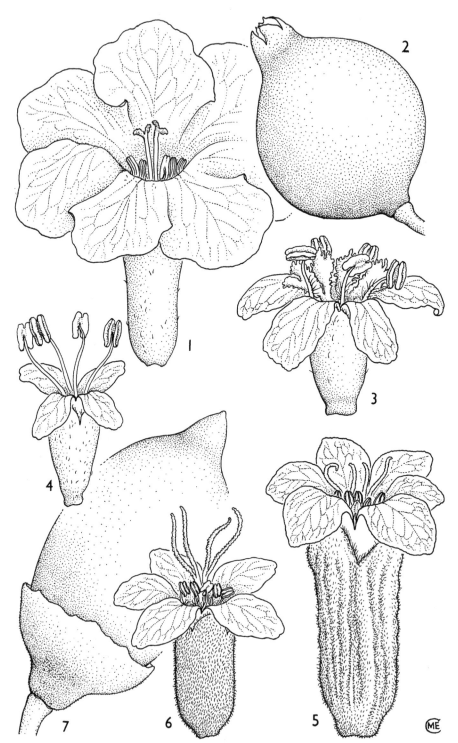

Fig. 1. *CORDIA SUBCORDATA*—**1**, flower, × 2; **2**, fruit, × 2. *C. MYXA*—**3**, ♀ flower, × 4. *C. PETERI*—**4**, ♂ flower, × 4. *C. FISSISTYLA*—**5**, flower, × 4. *C. MILLENII*—**6**, ♀ flower, × 4; **7**, fruit, × 2. 1, from *Greenway* 13141; 2, from *Greenway* 10441; 3, from *Ruffo* 1233; 4, from *Peter* 52440; 5, from *Ludanga* 1209; 6, 7, from *Chandler* 1151. Drawn by Mrs Maureen Church.

TANZANIA. Tanga District: Tanga–Pangani road junction with Kigombe–Dhali road, 16 Mar. 1955, *Faulkner* 1586!; Bagamoyo District: 3 km. SE. of Bagamoyo, 26 July 1970, *Thulin & Mhoro* 487!; Rufiji District: Mafia I., 31 Mar. 1933, *Wallace* 833!; Zanzibar I., Makunduchi, 4 Jan. 1960, *Faulkner* 2454!
DISTR. **K** 7; **T** 3, 6, 8; **Z**; **P**; Somalia (S.), Mozambique, Madagascar, Mascarene Is., Seychelles and Comoro Is., India, Indochina, Malaysia, Indonesia and Pacific Is.
HAB. Shore above high tide mark; inland fringe of mangrove swamps; littoral savanna; 0–1.5 (–75) m.

NOTE. Heine gives the synonymy, none of which concerns Africa. It is sometimes stated that this is the only species of subgen. *Cordia* which occurs naturally in the Old World but *C. suckertii* Chiov. is undoubtedly another.

2. **C. myxa** L., Sp. Pl.: 190 (1753); Del., Fl. Egypte: 47, t. 19 (1812); Boiss., Fl. Orient. 4: 124 (1879); Bak. & Wright in F.T.A. 4(2): 14 (1905); Hutch. in K.B. 1918: 217, fig. on p. 220 (1918); Dale, Introd. Trees Uganda: 24 (1953); Aubrév., Fl. For. Côte d'Ivoire, ed. 2, 3: 218, t. 334/6 (1959); E.P.A.: 768 (1961); Heine in F.W.T.A., ed. 2, 2: 320 (1963); Riedl in Fl. Iran. 48/15: 7 (1967); Taton in F.C.B., Borag.: 16, map. 4 (1971); Edmondson in Fl. Turkey 6: 246 (1978); Townsend in Fl. Iraq 4(2): 645 (1980); Meikle, Fl. Cyprus: 1120 (1985); Burkill, Useful Pl. W. Trop. Afr. 1: 287 (1985); Warfa, Acta Univ. Upsal. 174: 41, fig. 5/B, 8 (1988); Martins in F.Z. 7(4): 65 (1990). Type: ? Middle East, *Hasselquist* (S-LINN, fiche 94.5, lecto.)

Tree 6–12 m. tall, sometimes rather twisted, with young stems hairy but soon glabrous, older with circular petiolar scars. Leaves broadly ovate to subcircular, or sometimes obovate, 3–18 cm. long, 3–20 cm. wide, rounded to cordate or cuneate at the base, rounded to shortly obtusely acuminate at the apex, entire or repand-dentate, subcoriaceous, glabrous above, glabrous to ± densely pubescent beneath or even velvety; petiole 0.6–3.5 cm. long, glabrous or sparsely hairy. Cymes in terminal lax panicles, often on short lateral branches, 3–8.5 cm. long, 2–7 cm. wide, axes glabrous to sparsely pubescent; pedicels 1–2 mm. long, articulate at the apex. Male: calyx campanulate, 4.5–5.5 mm. long, 3-lobed, glabrous outside, pubescent to tomentose at apex inside; corolla white; tube 3.5–4.5 mm. long; lobes 5, elliptic, 5 mm. long, 2 mm. wide, reflexed; stamens exserted, the filaments 1.5–3.5 mm. long, hairy at the base; ovary rudimentary and style absent. Female: calyx tubular-campanulate, 6–8.5 mm. long, irregularly 3–4-toothed, glabrous outside save for tips of lobes, densely pubescent inside; corolla-tube 4.5–6.5 mm. long; lobes 4–6, elliptic to obovate, 5–7 mm. long, 2.5–3.5 mm. wide, reflexed and rolled up; stamens with filaments 1.5–2.5 mm. long, ± pubescent; anthers sterile; ovary ellipsoid or obovoid, 2.5–3.5 mm. long, 2–2.7 mm. wide; style exserted, 8–9 mm. long, deeply divided into 4 stigmatic branches 4–5 mm. long, flattened and subfoliaceous with irregular or erose-denticulate margins. Fruit yellow, apricot or blackish, ovoid, (1.2–)2–3.5 cm. long, apiculate, held in the accrescent campanulate calyx (0.7–1 cm. long, 1.2–1.5(–2) cm. wide), which is ± obscurely lobed or subtruncate; pulp mucilaginous and sweet; endocarp broadly ellipsoid or ± globose, ± 1.2 cm. long, 1 cm. wide, deeply rugose, 4-locular but only 1 seed developing. Fig. 1/3.

UGANDA. Acholi District: Gulu, Oct. 1948, *Dale* U.669!
TANZANIA. Tabora District: Rungwa Game Reserve, 26 Sept. 1967, *Shabani* 53!; Morogoro, Indian Cemetery, Sept. 1951, *Eggeling* 6292!; Uzaramo District: Dar es Salaam, Oyster Bay, 15 July 1967, *Harris* 710!; Zanzibar I., Chwaka, 22 Sept. 1962, *Faulkner* 3103!
DISTR. **U** 1; **T** 3, 4–6; **Z**; **P**; native of tropical Asia (India, etc.), cultivated and often naturalised in tropical Africa, Senegal to Cameroon, Zaire, Ethiopia, Malawi, Mozambique, Zimbabwe, Madagascar, also in N. Africa (Algeria, Libya, Egypt), Cyprus, Palestine, Arabia, Turkey, Iraq, S. Iran, Malaysia, Queensland, etc.
HAB. Naturalised in coastal and other bushland and cultivated; 0–1050 m.

SYN. *C. sp.* sensu U.O.P.Z.: 210 (1949)

NOTE. Much used for pickles but see Townsend and Burkill for other uses.
    Formerly much confused with *C. dichotoma* Forst.f. (*C. obliqua* Hutch. *non* Willd.) which differs in more elongate leaves and filiform stigmatic branches. The figure given by Heine under that name (Fl. Nouv.-Caléd. 7: 102, t. 22/4 & 6 (1976)) shows stigmas more like those of *C. myxa*. Johnston (Journ. Arn. Arb. 32: 10 (1951)) hints that the two are not so clearly distinct as Hutchinson claimed, a view with which Heine (in litt.) concurs. Warfa includes *C. obliqua* in his synonymy of *C. myxa* which judging by a leaf from Willdenow's original type sent to Kew by K. Schumann is probably correct.

3. **C. goetzei** *Gürke* in E.J. 28: 307, 461 (1900); Bak. & Wright in F.T.A. 4(2): 14 (1905); Senni, Gli alberi Formazioni legnose della Somalia: 211 (1935); T.T.C.L.: 76 (1949); E.P.A.: 768 (1961); K.T.S.: 70 (1961); Haerdi in Acta Trop., Suppl. 8: 150 (1964); Vollesen in Opera Bot. 59: 75 (1980); Warfa, Acta Univ. Upsal. 174: 39, fig. 3/C, D, 5/B (1988); Martins in F.Z. 7(4): 62, t. 20 (1990). Type: Tanzania, Kilosa/Morogoro District, between Kisaki and the Ruaha R., *Goetze* 365 (B, holo. †, BR, K, iso.!)

Erect shrub or tree 2–15(–18) m. tall, with a peculiar angular vertically ridged trunk ('as if round originally and has then collapsed'), variously described as 'fluted' and 'spiralig gedrecht' (Scheffler); branches sometimes leaning and almost scandent; bark usually grey-white and smooth (as if white-washed or painted (Greenway)), peeling off to reveal yellow-green or green surface beneath but occasionally said to be rough (see note); young branches glabrous (see note), longitudinally ridged when dry. Leaf-blades elliptic, (1.5–)2–12.5(–13.5) cm. long, 1–6.5 cm. wide, rounded to ± acute at the apex, cuneate to rounded at the base, entire, quite smooth and glabrous or with dense minute tubercles above (see note); venation slightly raised and finely reticulate beneath; petiole (1–)1.5–3 cm. long. Flowers sweet-scented in lax cymes, terminal (usually on lateral shoots) or axillary; peduncle 0.2–1.5 cm. long; pedicels 0.5–4 mm. long. Male: calyx narrowly campanulate, pubescent inside; tube 3 mm. long, 3-lobed, the lobes obtuse, 1 mm. long; corolla white or yellowish; tube 3–4 mm. long; lobes 4–5, narrowly oblong-lanceolate to obovate-oblong, (2.5–)4.5–5 mm. long, 2 mm. wide; stamens well exserted; filaments (2–)4–5 mm. long with a few hairs on the adnate parts; rudimentary ovary 0.7 mm. diameter; style absent. Female: calyx more cylindrical, pubescent inside; tube 3 mm. long, not ribbed; lobes 4, ovate, ± with apical tufts of hair; corolla-tube cylindrical, swollen at the base, 3–4 mm. long; lobes oblong-elliptic, 4 mm. long, 1–2 mm. wide; stamens with upper free parts of filaments 1 mm. long, ovary ellipsoid, 2 mm. long, 1.2 mm. diameter; unbranched part of style 2–3 mm. long; primary branches 1 mm. long with 4 stigmatic arms 2.2 mm. long. Fruiting inflorescence up to 4 cm. long; fruits subglobose to obovoid, 0.8–1(–1.9) cm. long, 6(–13) mm. wide; endocarps woody, 4–8(–11) mm. long, 4–6.5(–8) mm. wide; fruiting calyx cupular, 5 mm. long, 7 mm. wide at throat, ± constricted below. Fig. 2.

KENYA. Masai District: Tsavo National Park, Mzima Springs, 17 Jan. 1961, *Greenway* 9757!; Kilifi District: Vipingo, 20 Dec. 1953, *Verdcourt* 1115!; Tana River District: Bura, 7 Mar. 1963, *Thairu* 46!
TANZANIA. Mbulu District: Lake Manyara National Park, near Main Gate, 24 Feb. 1964, *Greenway & Kanuri* 11228!; Lushoto District: Mombo Forest Reserve, 24 Dec. 1962, *Semsei* 3625!; Morogoro District: 1.6 km. S. of Mkata Railway Station, 28 Jan. 1957, *Welch* 364 ♀! & 366 ♂!; Zanzibar I., Mkokotoni, 18 Dec. 1963, *Faulkner* 3328!
DISTR. **K**4, 6, 7; **T**1–4, 6–8; **Z**; Somalia (S.), Mozambique, Zambia, Malawi, Zimbabwe
HAB. Evergreen coastal forest, rain-forest, riverine forest and thicket ground-water forest; 0–1350 m.

SYN. *C. ravae* Chiov., Fl. Somala 2: 309, figs. 179, 180 (1932); Senni, Gli alberi formazioni legnose della Somalia: 211, fig. 75 (1935); E.P.A.: 769 (1961). Type: Somalia (S.), R. Juba, Touata I., Alessandra, *Tozzi* 332 (FT, holo, K, iso.!)
　　[*C. myxa* sensu Bak. & Wright in F.T.A. 4(2): 14 (1905), quoad *Kirk* (Mozambique), *Scott* (Malawi) & *Menyharth* 866, non L.]

NOTE. Descriptions of the trunk and bark vary. My own recollection is of a smooth grey-white bark and curiously angular-fluted rather twisted trunk which is very characteristic. Tanner's description of the bark as rough must I think be erroneous; Welch and others describe it as liver-coloured and brown. Although well characterized in the Flora area problems arise further south, there seeming to be some introgression with a very diverse species with narrow subsessile often hairy leaves which grows on termite mounds, *C. mukuensis* Taton (*Ehretia guerkeana* De Wild., non *Cordia guerkeana* Loes.) — see Taton in B.J.B.B. 41: 258 (1971) & F.C.B., Borag.: 18 (1971); Martins in F.Z. 7(4): 64 (1990) — type: Zaire, Shaba, Lubumbashi, Mt. Mukuen, *Schmitz* 2654 (BR, syn.!). Taton synonymised *Ehretia guerkeana* with *C. senegalensis* Juss. but I agree with P. Bamps (*in litt.*) this is incorrect. These 3 species are likely to trouble a future monographer. *Greenway* 9038 (Musoma District, Grumeti R. Game corridor, 9 Nov. 1956) first thought to be an undescribed species is only a small-leaved variant of *C. goetzei*.

4. **C. peteri** *Verdc.* sp. nov., ob faciem et indumentum simile probabiliter *C. stuhlmannii* Gürke affinis sed foliis latioribus oblongioribus minus acuminatis, petiolis crassioribus supra manifeste canaliculatis, indumento surculorum juvenilium stricte appresso haud denso differt. Typus: Tanzania, E. Usambara Mts., Sigi–Longuza [Longusa], *Peter* 52440 (B, holo.!)

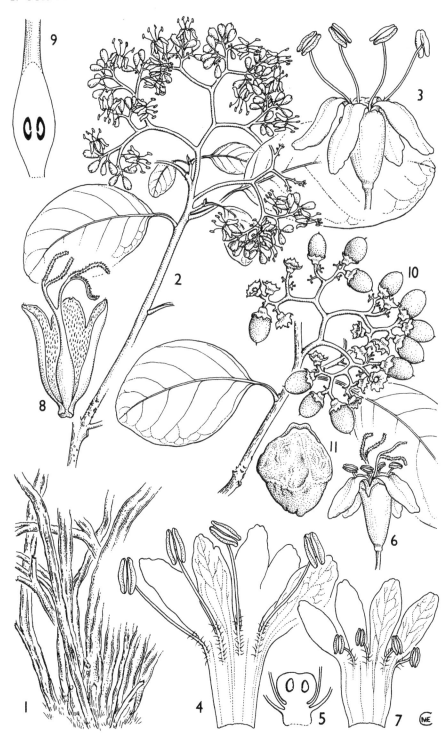

FIG. 2. *CORDIA GOETZEI*—**1**, several-stemmed base of tree; **2**, ♂ flowering branch × ⅔; **3**, ♂ flower, × 4; **4**, same, corolla opened out, × 5; **5**, rudimentary ovary, longitudinal section, × 16; **6**, ♀ flower, × 4; **7**, same, corolla opened out, × 5; **8**, calyx, opened out to show ovary, × 6; **9**, ovary, longitudinal section, × 12; **10**, fruiting branch, × ⅔; **11**, pyrene, × 3. 1, from *Milne-Redhead & Taylor* 7068; 2–5, from *Greenway & Kanuri* 11228; 6–9, from *Tanner* 1722; 10, 11, from *Richards* 18307. Drawn by Mrs Maureen Church.

Habit unknown, presumably a shrub or tree; young stems fairly densely closely adpressed pubescent with pale hairs but soon glabrous, ± ridged, lenticellate and roughened with nodular petiole-bases. Leaf-blades oblong-elliptic, 4–13 cm. long, 1.8–6 cm. wide, acute or shortly acuminate at the apex, ± rounded or cuneate at the base, sparsely pubescent with pale adpressed hairs, later almost glabrous save for scattered hairs above and on nerves beneath, ± discolorous; main lateral nerves rather few, 3–5 on each side; petiole 1.2–2.5 cm. long. Inflorescences axillary from upper axils, paniculate, ± 10 cm. long but probably more extensive; peduncle 2.5 cm. long; secondary branches 2.5–5 cm. long; ultimate branches 5 mm.; pedicels ± 1.5 mm.; all axes and outside of calyx with similar pubescence to the stems. Male flowers: calyx-tube obconic, ± 5 mm. long, densely silky pilose inside; lobes ± triangular, 0.5–1 mm. long; corolla white, tube very narrowly funnel-shaped, 5 mm. long, glabrous outside, slightly pilose at the throat; lobes elliptic-oblong, 3.2–3.5 mm. long, 2–2.4 mm. wide, obtuse; filaments exserted 3.5–5.5 mm.; rudimentary ovary 0.8 mm. diameter, ± 4-lobed with few erect hairs at tip; style absent. Female flowers and fruits not seen. Fig. 1/4, p. 8.

TANZANIA.   Lushoto District: E. Usambara Mts., Sigi–Longuza, 17 May 1917, *Peter* 52440!
DISTR.   T3; known only from the type gathering
HAB.   Lowland evergreen forest; 500 m.

NOTE.   It is surprising that only this specimen appears to have been collected in an area only a mile or so from a centre of botanical collecting, but much of the lowland forest was cleared very early.

5. **C. fissistyla** *Vollesen* in Nordic Journ. Bot. 1: 325, fig. 1 (1981). Type: Tanzania, Kilwa District, Selous Game Reserve, Kingupira, *Vollesen* in *MRC* 2283 (C, holo., DSM!, EA!, K!, WAG, iso.)

Deciduous, dioecious, erect or scandent shrub to ± 4 m. tall; young branches 4-angled, with dense yellowish white velvety indumentum interspersed with scattered long hairs, becoming rounded, pale brown, glabrous and with scattered lenticels. Leaf-blades broadly elliptic, obovate, round or transversely elliptic, 4.5–15.5 cm. long, 3.5–14 cm. wide, rounded to retuse at the apex but with a central broad short acumen, attenuate, truncate or subcordate at the base, the margin coarsely irregularly toothed but less so towards the base, densely velvety with similar indumentum to branches when young, becoming less hairy with age, sparsely pilose above but still ± velvety beneath; venation raised and reticulate beneath; petiole 0.8–10.5 cm. long. Inflorescence a few-flowered (5–10 or up to 50 on sucker shoots) terminal panicle of indistinctly scorpioid cymes; peduncle 0.6–1.2 cm. long, elongating in fruit; cyme-branches similar; pedicels obsolete; all axes with indumentum similar to that of young branches. Male flowers: calyx funnel-shaped, 4–8 mm. long with 3–4 triangular acute teeth 2–5 mm. long, longitudinally ribbed outside, densely whitish velvety outside, densely sericeous inside with glossy hairs; corolla yellowish white, tube funnel-shaped, longitudinally ribbed, particularly at base, 0.9–1.3 cm. long; lobes 5–6, triangular, 3–5 mm. long, 5 mm. wide; stamens 5–6; free part of filaments 5 mm. long, adnate part hairy; rudimentary ovary absent. Female flowers: calyx urceolate, 1.2–2 cm. long, with 3–4 triangular acute 3–5 mm. long teeth, with similar indumentum to ♂ flowers; corolla same colour as ♂, funnel-shaped with inflated base, 1.2–1.6 cm. long, obscurely ribbed; lobes 5–7, triangular, 4–5 mm. long and wide; sterile stamens up to 7, with empty anthers, otherwise as in ♂; ovary ovoid, ± 1 cm. long including a ± 2 mm. long beak, silky, with 4 free styles or rarely fused in pairs for 1–2 mm., ± 1 cm. long including the 6 mm. long stigmas. Fruit brown, ellipsoid, 2.5–3 cm. long, glabrous save for apical 5 mm. and the beak; cupuliform fruiting calyx 1.8–2.4 cm. long, distinctly reticulately nerved, ± velvety. Fig. 1/5, p. 8.

TANZANIA.   Kilwa District: Ngarambe Ujamaa village, 28 Sept. 1970, *Ludanga* in *MRC* 1147! & same area, 4 Feb. 1971, *Ludanga* in *MRC* 1209! & Selous Game Reserve, Kingupira, 27 Feb. 1977, *Vollesen* in *MRC* 4493!
DISTR.   T 8; not known elsewhere
HAB.   On old termite mounds with dense thicket and some trees, at edges of temporary water-holes and at edges of riverine thickets, always on very alkaline soil; 100–125 m.

SYN.   *C. sp. nov.* sensu Vollesen in Opera Bot. 59: 75 (1980)

NOTE.   A distinct species of no close affinity occurring in a very restricted area. Vollesen suggests it is possibly related to *C. pilosissima* Bak.

6. **C. millenii** *Bak.* in K.B. 1894: 27 (1894); Bak. & Wright in F.T.A. 4(2): 11 (1905); E.A. Bruce in K.B. 1940: 62 (1940); F.P.N.A. 2: 127 (1947); I.T.U., ed. 2: 48 (1952); F.P.S. 3: 77 (1956); Aubrév., Fl. For. Côte d'Ivoire, ed. 2, 3: 224 (1959); K.T.S.: 70 (1961); Heine in F.W.T.A., ed. 2, 2: 320 (1963); Taton, F.C.B., Borag.: 10, t. 1 (1971); Burkill, Useful Pl. W. Trop. Afr. 1: 287 (1985); Warfa, Acta Univ. Upsal. 174: 36 (1988). Type: Sierra Leone, Lagos, *Millen* (K, holo.!)

Shrub to tall tree, 4–32 m. tall, with spreading crown and thick ± short rarely straight bole, 9–12 m. long, bearing short blunt buttresses; bark pale brown, greenish grey or grey-brown, rough and fibrous, flaking in oblong pieces; slash pale yellow turning blackish orange or pale creamy brown, fibrous and layered, turning rapidly through greenish to dark brown; young shoots ferruginous-pubescent. Leaf-blades rounded-obovate, ovate-elliptic to almost round, (3.5–)9–15(–33) cm. long, (2–)6–11.5(–26) cm. wide, ± rounded to acuminate at the apex, rounded to cordate at the base, entire or crenulate, scabrid or glabrous above, tomentose-pilose or softly velvety pubescent beneath or hairy only on the venation which is raised and reticulate beneath; petiole 2.5–7.5(–16) cm. long, ferruginous-pilose, leaving marked scars after leaf-fall. Cymes condensed, in large panicles 20–45 cm. long; peduncles to 17 cm. long, secondary axes up to 4 cm. long; pedicels short or obsolete, articulate. Flowers unisexual, very sweetly scented. Male: calyx tubular-campanulate, 0.8–1.2 cm. long, 3–4-toothed or ± 2-lipped, fulvous-tomentose, ± ribbed; corolla white, yellow or greenish yellow, or cream; tube cylindric-campanulate, 0.7–1.2 cm. long; limb 1.2–1.5 cm. wide, the lobes 5–7, oblong, elliptic or narrowly ovate, 4.5–7 mm. long, 1.8–2.5 mm. wide; stamens 5–7, exserted, with filaments 5–9 mm. long; rudimentary ovary subglobose, 0.5–0.9 mm. diameter, with an apical tuft of hairs. Female: calyx more tubular, 6–10 mm. long, 3–4-lobed; corolla-tube 5–10(–12) mm. long; limb with lobes similar to ♂ but elliptic-oblong to ovate, 4.5–6.5 mm. long, 2.5–4 mm. wide; stamens 5A2—7, scarcely exserted, the pilose filaments 2.5–3.5 mm. long, sterile; ovary ovoid, 3.5 mm. long, 2 mm. diameter, glabrous; style 5 mm. long with 2 branches 3 mm. long, each with stigmatic branches linear, flattened 4–6 mm. long, fimbriate, exserted. Fruit obovoid, ellipsoid or ovoid, 2–4.4 cm. long, 1.4–3 cm. diameter, shortly apiculate, sitting in an enlarged calycine cupule 1.7–2 cm. long, 2.5 cm. wide; putamen subquadrate, grooved, with 1–2 locules developed; seeds 1–2. Figs. 1/6,7, p. 8; 3, p. 14.

UGANDA. Bunyoro District: Budongo Forest, Mar. 1933, *Eggeling* 1155!; Masaka District: Sese Is., Bugala Entebbe Botanic Gardens, Mar. 1934, *Chandler* 1151!
KENYA. N. Kavirondo District: Kakamega Forest, Mar. 1934, *Makin* in F.D. 3284! & Oct. 1950, *Forest Dept.*, comm. *Cobby* in Bally 8208! & same area, Ikuywa R., 8 km. ESE. of Kisieni, 6 Jan. 1968, *Perdue & Kibuwa* 9469!
TANZANIA. Kigoma District: Ititie, 26 Dec. 1963, *Azuma* 1017!; Mpanda District: Mahali Mts., *Nishida* 8! & same area, Kasoje, 24 Sept. 1958, *Newbould & Jefford* 2625!
DISTR. U 1*, 2, 4; K 5; T 4; Guinée to Gabon and Angola across to the Central African Republic, Zaire and Sudan
HAB. Rain-forest; 900–1650 m. (see note)
SYN. *C. longipes* Bak. in K.B. 1894: 27 (1894); Hiern, Cat. Afr. Pl. Welw. 1: 714 (1898); Bak. & Wright in F.T.A. 4(2): 11 (1905). Type: Angola, Sobato de Bumba, *Welwitsch* 5428 (K, holo.!, BM, iso.!)
  C. chrysocarpa Bak. in K.B. 1894: 27 (1894); Hiern, Cat. Afr. Pl. Welw. 1: 713 (1898); Bak. & Wright in F.T.A. 4(2): 11 (1905). Type: Angola, Golungo Alto, ascent to Capopa, *Welwitsch* 5461 (K, holo.!, BM, iso.!)
  C. irvingii Bak. in K.B. 1895: 113 (1895), pro parte excl. syntypo *Irving*; Bak. & Wright in F.T.A. 4(2): 12 (1905). Type: Nigeria, 'Western Lagos', *Rowland* (K, lecto.!)
  C. liebrechtsiana De Wild. & Th. Dur. in Compte Rendu Soc. Bot. Belge 38: 38 (1899); Bak. & Wright in F.T.A. 4(2): 13 (1905). Zaire: without locality, *Dewèvre* (BR, holo.)
  C. unyorensis Stapf in J.L.S. 37: 527 (1906), pro parte, fruit only. Type: "Unyoro, Bugoma & Budongo Forest", *Dawe* 798 (K, syn.!)
  C. ugandensis S. Moore in J.B. 54: 288 (1916). Type: Uganda, Mengo District, Kijude, *Dummer* 2727 (BM, holo.!, K, iso.!)
NOTE. A well-known timber tree. "Wood yellow to yellow-brown, lustrous, seasoning and finishing well, very suitable for furniture and cabinet making" — "the tree is a favourite for dug-out canoes both because it is easily adzed and floats if overturned — also used for making drums" (I.T.U., ed. 2). Cufodontis (E.P.A.: 768 (1961)) states that Breitenbach (1960) records this from Ethiopia, but it is not mentioned in his Indig. Trees Ethiopia, ed. 2 (1963). The altitudinal range is not known, hardly a collector mentioning it.

* Madi, Zoka Forest, *fide* I.T.U.

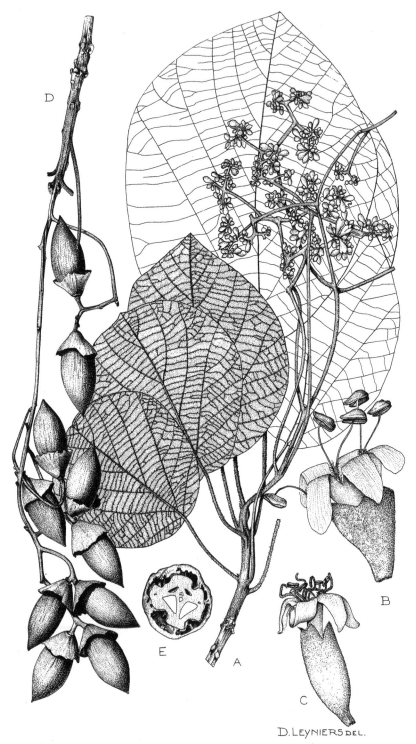

D. Leyniers del.

Fig. 3. *CORDIA MILLENII*—**A**, flowering branch, × ½; **B**, ♂ flower, × 3; **C**, ♀ flower, × 3; **D**, infructescence, × ½; **E**, fruit, transverse section, × 1. A, from *Ghesquière* 4409; B, from *Luja* 3; C from *Ghesquière* 1156; D, from *Louis* 11289. Drawn by D. Leyniers. Reproduced with permission from Flore du Congo du Rwanda et du Burundi.

7. **C. uncinulata** *De Wild.* in Rev. Zool. Afr. 9(3), Suppl. Bot.: B 89 (1921) & in Pl. Bequaert. 2: 119 (1923); F.P.N.A. 2: 127 (1947); Taton, F.C.B., Borag.: 15, map 4 (1971); Warfa, Acta Univ. Upsal. 174: 47 (1988). Lectotype chosen by Taton (1971): Zaire, Kivu, Lesse–Beni, *Bequaert* 3261 (BR, lecto.! & isolecto.!)

Shrub or woody liane 1.8–15 m. tall or long; young branches brown, striate, with pale ferruginous and white hairs, later grey, obscurely lenticellate and glabrous, with short recurved spurs which are accrescent petiole-bases ± 3 mm. long, eventually subtending axillary shoots long after the leaves have fallen. Leaf-blades elliptic to ovate-elliptic, 2.5–13 cm. long, 2.5–8 cm. wide, acuminate at the apex, cuneate, rounded or ± subcordate at the base, the margins entire to ± crenate in upper part, ± thin, discolorous, very shortly sparsely pubescent on both surfaces, not scabrid; petiole 0.5–5 cm. long. Cymes terminal, dichotomous; peduncle and main axes 1–1.5 cm. long, ferruginous pubescent; pedicels 1–2 mm. long, articulate, pubescent. Male: calyx tubular-campanulate, 5–7 mm. long, sparsely pubescent outside, glabrous inside, with 4–7 teeth; corolla white; tube 4–5 mm. long; lobes 4–5, elliptic to narrowly obovate or narrowly oblong, 4–6 mm. long, 1.3–2 mm. wide; stamens exserted, the filaments 5.5–6.5 mm. long, hairy at the base; anthers 1.5 mm. long; rudimentary ovary minute, under 1 mm. long; style absent. Female flowers similar but filaments 1–1.5 mm. long with a few hairs at the base; anthers sterile, 0.6 mm. long; ovary ovoid, 4 mm. long, 3 mm. diameter; style 1.5 mm. long; primary branches 1.5–2 mm. long, with stigmatic branches 3.5 mm. long. Fruit orange, ovoid, ± 1 cm. long, 5–6 mm. diameter, usually with 2 locules, tipped by style-base; cup shallow, 1–1.5 cm. wide, 7–10 mm. long. Fig. 4/1, p.16.

UGANDA. Bunyoro District: Budongo Forest, *Sangster* 167!; Mengo District: km. 16 on the Entebbe road, Kajansi Forest, Nov. 1937, *Chandler* 2017!
DISTR. U 2, 4; Senegal, Nigeria, Zaire, Central African Republic, probably Sudan, Ethiopia
HAB. Open evergreen forest; ± 1200 m.

SYN. *C. tisserantii* Aubrév., Fl. For. Soud.-Guin.: 490, 493, t. 110/1 (1950); Burkill, Useful Pl. W. Trop. Afr. 1: 287 (1985), *nom. invalid.*, French description only, Central African Republic, Moroubas, *Tisserant* 587 (P) & Ippy, *Tisserant* 1871 (P) & 50 km. N. of Bambari, *Tisserant* 1900 (P, BM!) & S. Nigeria, *Foster* 325 (K!, ?P)

NOTE. It is remarkable that this has not been recollected seeing that it occurs or did so on the short road joining the capital with the airport. *Gérard* 5292 (Zaire, Bambesa) has thicker leaves and fruits to 1.3 cm. long.

8. **C. monoica** *Roxb.*, Pl. Coromandel 1: 43, t. 58 (1796) & Fl. Indica, ed Carey 2: 334 (1824); DC., Prodr. 9: 479 (1845); C.B. Cl. in Fl. Brit. Ind. 4: 137 (1883); K. Mathew, Fl. Tamilnadu Carnatic 1: 268 (1981) & 3: 999 (1983); Warfa, Acta Univ. Upsal. 174, fig. 5/A, 6, 7/A (1988); Verdc. in K.B. 44: 166 (1989); Martins in F.Z. 7(4): 65 (1990). Type: India, E. Madras, *Roxburgh* drawing No. 200 (K, lecto.!)

Much-branched spreading shrub or small tree branched from the base, 1.5–8(–15) m. tall; crown usually spreading; bark rather rough, grey, but sometimes smooth and flaking in long strips (see note); branches longitudinally striate, with ± ferruginous indumentum and often velvety when young, later glabrescent, with short internodes and prominent reniform or semicircular petiole-scars. Leaves alternate; blades elliptic, ovate, obovate or ± rounded, (1–)2.5–8(–11 or –13 on coppice) cm. long, (1–)2–9 cm. wide, rounded, apiculate or often emarginate at the apex, less often acute, cuneate to rounded at the base, entire, denticulate or rarely distinctly repand-dentate, rugose and mostly very scabrid above with minute tubercle-based hairs together with sparse branched hairs, densely ± softly white pubescent beneath; venation reticulate beneath, often reddish; petiole (0.5–)1.2–5 cm. long, brownish tomentose. Flowers fragrant, ♂ or unisexual in terminal or axillary panicles of cymes 2–6 cm. long, few-flowered, the axes fulvous-tomentose; pedicels articulate at the apex, 0.2–0.8(–3) mm. long. Calyx tubular-campanulate, 6–8.5 mm. long, 5–8 mm. wide, tomentose outside with simple and branched hairs, glabrous or sparsely pubescent inside, 3–5-lobed or sometimes ± 2-lipped or irregularly torn. Corolla greenish white to pale yellow, 1–1.2 cm. wide; tube cylindric, 4–7 mm. long; lobes 3–5, oblong to narrowly obovate, 3.5–5.5(–6.5) mm. long, 1–2.5 mm. wide, spreading. Male: filaments either 6 mm. long or 2–3 mm. long in what appears to be different forms; corolla and stamens 5(–6)-merous; gynoecium absent. Female and ♂: filaments 2–3 mm. long; corolla 4–5-merous; ovary ovoid, 3–3.5(–5) mm. long, 1.5–1.8 mm. wide, glabrous; style exserted, 6–10 mm. long; stigmatic branches 3–5 mm. long, recurved. Fruits yellow or orange but also reported to be olive-black to black, ovoid or obovoid, 0.9–2 cm. long,

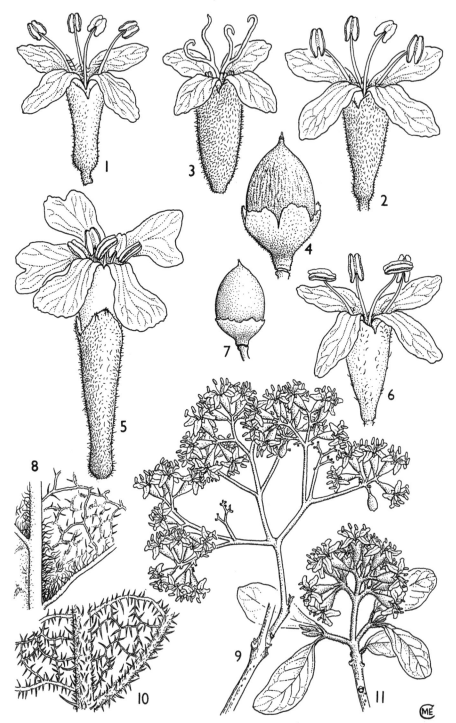

FIG. 4. *CORDIA UNCINULATA*—**1**, ♂ flower, × 4. *C. MONOICA*—**2**, ♂ flower, × 4; **3**, ♀ flower, × 4; **4**, fruit, × 2. *C. ELLENBECKII*—**5**, ♂ flower, × 4. *C. CRENATA* subsp. *MERIDIONALIS*—**6**, ♂ flower, × 4; **7**, fruit, × 2. *C. SINENSIS*—**8**, part of lower leaf surface, × 10; **9**, inflorescence, × 1. *C. QUERCIFOLIA*—**10**, part of lower leaf surface, × 10; **11**, inflorescence, × 1. 1, from *Sangster* 167; 2, from *Gillett* 20066; 3, from *Brenan, Gillett & Kanuri* 14731; 4, from *Katende* 426; 5 from *Gillett* 12674; 6, from *Adamson* 46; 7, from *Synnott* 1732; 8, from *Gillett* 12571; 9, from *Fratkin* 2; 10, from *Gillett* 12706; 11, from *Newbould* 2954. Drawn by Mrs Maureen Church.

0.6–1.2 cm. wide, sitting in the persistent 1 cm. wide accrescent calyx, 3–4-locular but usually 1-seeded by abortion. Fig. 4/2–4.

UGANDA. Karamoja District: Moroto, 2 Jan. 1937, *A.S. Thomas* 2133!; Bunyoro District: Butiaba Escarpment, May 1937, *Eggeling* 3343!; Teso District: Serere, Mar. 1932, *Chandler* 611!
KENYA. Northern Frontier Province: Samburu, Uaraguess [Warges], 30 Nov. 1958, *Newbould* 2976!; Nairobi, Mbagathi, Feb. 1930, *Dale* in *F.D.* 2512!; Masai District: Ewaso Ngiro–Loliondo road where it crosses the Masandari [Masan] R., Subatai [Subatia], 14 Dec. 1963, *Verdcourt* 3839!
TANZANIA. Arusha District: Ngurdoto Crater National Park, Small Momella Lake, 2 Mar. 1966, *Greenway & Kanuri* 12415!; Kondoa, 13 Mar. 1928, *B.D. Burtt* 1518!; Morogoro, Mar. 1955, *Semsei* 1958!; Zanzibar I., Kizimkazi, 10 Jan. 1931, *Vaughan* 1809!
DISTR. U 1–4; K 1–7; T 1–3, 5–8; Z; P; Zaire, Burundi, Somalia, Sudan, Ethiopia, Mozambique, Zimbabwe, Botswana, Angola, South Africa and Namibia and also in India and Ceylon and cultivated in Mauritius
HAB. Very catholic, from quite wet evergreen forest to *Acacia* woodland, *Acacia-Commiphora* bushland and *Acacia-Euphorbia* thicket in grassland, coastal thicket, etc., often riverine; 0–1825 m.

SYN. *C. dioica* DC. & A.DC. in DC., Prodr. 9: 481 (1845), probabiliter et certe sensu Vatke in Linnaea 43: 315 (1882) et Bak. & Wright in F.T.A. 4(2): 16 (1905), pro parte. Type: Pemba I., *Bojer* (G, lecto.)*
C. ovalis DC. & A.DC. in DC., Prodr. 9: 479 (1845); Gürke in E. & P. Pf. IV. 3A: 83 (1893); Bak. & Wright in F.T.A. 4(2): 15 (1905); R.E. Fries, Wiss. Ergebn. Schwed. Rhod.-Kongo Exped. 1: 271 (1916); Robyns, F.P.N.A. 2: 127 (1947); T.T.C.L.: 76 (1949); I.T.U., ed. 2: 49 (1952); K.T.S.: 70 (1961); E.P.A.: 768 (1961); Taton in F.C.B., Borag.: 13, map 4 (1971); Vollesen in Opera Bot. 59: 75 (1980). Type: Ethiopia, Gafta [& Guendepta], *Schimper* 1218 (G, lecto., K, isolecto.!)
[*C. myxa* sensu A.Rich., Tent. Fl. Abyss. 2: 82 (1850), pro parte quoad *Schimper* 1582, non L.]
C. rubra Hiern, Cat. Afr. Pl. Welw. 1: 712 (1898). Types: Angola, near Barra do Dande to mouth of R. Cuanza, *Welwitsch* 5423 (LISU, syn., BM, isosyn.!), Golungo Alto, *Welwitsch* 5431 (LISU, syn., BM, isosyn.!) & Bumbo, *Welwitsch* 4783 (LISU, syn., BM, isosyn.!) & Ethiopia, Gafta, *Schimper* 1582 (BM, syn., K, P, etc., isosyn.!)
C. obovata Bak. in K.B. 1894: 28 (1894), *non* Balf.f., *nom. illegit.* Type: Ethiopia, Agow, Dscha dscha, *Schimper* 2180 (K, lecto.!)
C. bakeri Britten in J.B. 33: 88 (1895). Type as for *C. obovata*
C. quarensis Gürke in P.O.A. C: 335 (1895) & in E.J. 28: 308 (1900); Merker in N.B.G.B. 3: 197 (1902); Bak. & Wright in F.T.A. 4(2): 18 (1905). Type: Tanzania, Kilimanjaro, Kware [Quare] stream, *Volkens* 2040 (B, syn.†, BM, isosyn.!) & *Volkens* 2045 (B, syn.†) & Mwanza District, Uzinza [Uzinja], Usambiro, *Stuhlmann* 845 (B, syn.†)
[*C. crenata* sensu Bak. & Wright in F.T.A. 4(2): 17 (1905); Broun & Massey, Fl. Sudan: 303 (1929); F.P.S. 3: 78 (1956), pro parte, *non* Del.]
C. bequaertii De Wild. in Rev. Zool. Afr. 9(3), Suppl. Bot.: B88 (1921), *non* De Wild. (1920). Type: Zaire, Kivu, Kabare, *Bequaert* 5432 (BR, holo.!)
C. kabarensis De Wild., Pl. Bequaert. 2: 118 (1923). Type as for *C. bequaertii*

NOTE. A very variable species with leaves generally so rough that they were used in the same way as sandpaper by native carpenters; the fruits are edible. Vatke (Linnaea 43: 315 (1882)), commented on the dioecious nature of the species. Small-leaved specimens with leaves to 3 × 2.5 cm. (e.g. *Richards & Arasululu* 26583 from Arusha District, Ngare Nanyuki, Ngaserai Plains and *Richards* 21601 from Lake Manyara National Park, near Neale R.) are distinctive but no more than ecotypes. *Greenway & Kanuri* 11079 (Lake Manyara National Park) has the leaves very sparsely pubescent beneath and calyx with much sparser pubescence than usual. Three specimens from Tanzania, Pare/Lushoto Districts (Buiko–Hedaru, 13 June 1915, *Peter* 52391, 600 m., Mkomazi [Mkomasi]–Mkumbara, 3 June 1915, *Peter* 52385, 480 m., and E. of Buiko, 1 June 1915, *Peter* 52379, 560 m.) have the undersurface of the leaves either ± glabrous or rough with scattered bulbous-based hairs. The note to *Verdcourt* 3895B (Kenya, Teita District, Voi, Mzinga Hill, 11 Jan. 1964) states "multistemmed bush collected to show the angular grey-white stem like *ravae*"; *Robertson & Luke* 5601 (Kenya, Lamu District, Boni Forest Reserve, 7.7 km. W. of Mararani [Marereni], 29 Nov. 1988) has very large leaves to 15.5 × 9.8 cm. and similar smooth peeling bark.
    Greenway 4282 (Tanzania, Masai District, Ketumbeine [Ketumbane] Mt., 7 Jan. 1936) had been determined as *C. quarensis*. Although I think it is merely a form of *C. ovalis* the calyx is about 1.1 cm. long including rounded 1.5 mm. long lobes, with a ± smooth basal part (due to developing ovary) contracted below and a cylindrical ± ribbed upper part with white hairs and ferruginous tomentum outside particularly above; possibly a specimen with prematurely accrescent calyx (see also *C. fischeri*, p. 18).
    Despite scarcely scabrid or ± non-scabrid upper leaf-surfaces I think *Koritschoner* 1821 (Shinyanga) and *Gillett & Glover* 19635 (Teita District, 10 km. W. of Bura Station, 2 Apr. 1972) and perhaps several other doubtful specimens are probably forms of *C. monoica* rather than *C. crenata* subsp. *meridionalis*, although the possibility of hybridization should not be overlooked.
    Interestingly I.T.U., ed. 2: 49 (1952) states "all Uganda material originally referred to *Cordia monoica* Roxb. is included here" and Eggeling (I.T.U.: 23 (1940)) gives the name in synonymy.

* A.DC. also cites a Bojer specimen from Madagascar which he had not seen but is nevertheless strictly speaking a syntype.

9. **C. fischeri** *Gürke* in P.O.A. C: 335 (1895) & in E.J. 28: 307 (1900); Bak. & Wright in F.T.A. 4(2): 10 (1905); T.T.C.L.: 76 (1949). Type: S. Kenya/N. Tanzania, Masai Highlands, Massaini, *Fischer* 437 (B, holo. †)

Tree or shrub, the branches pubescent above. Leaf-blades 9 cm. long, 6 cm. wide, obtuse at the apex, entire, coriaceous, rough above, densely pubescent beneath; petiole 1–3 cm. long. Cymes few-flowered ('cyma pauciflora, floribus ad apices ramorum 3–4 ...') at apices of branchlets, very shortly pedunculate or sessile. Calyx turbinate-tubular, 8 mm. long, 4–5-toothed, obsoletely 10-grooved, subtomentose outside. Corolla-tube rather slender, 1½ times (but also says 'ist nur wenig langer') as long as the calyx; ovate lobes reflexed downwards. Stamens slightly exserted, the 1 mm. long anthers on 1 mm. long filaments. Ovary ovate, 3 mm. long, acute, glabrous; style divided from the base the branches 7–9 mm. long, glabrous and split about halfway for a second time.

KENYA./TANZANIA.   Masai District: Massaini, *Fischer* 437
DISTR. **K6/T2**
HAB.   Unknown, presumably dry bushland

NOTE.   Gürke puts *C. fischeri* in sect. *Gerascanthus* sensu ampl. Gürke, i.e. with *C. africana* from his point of view at that date, which would rule out the possibility it was the same as *Greenway* 4282 (see p. 17) which I had considered a distinct probability. Fischer passed close to Ketumbeine Mt. and the description agrees to some extent. The leaves of *C. fischeri* are a good deal larger and the style differently divided if Gürke's description is taken at face value. The rough leaves and few-flowered inflorescences militate against any relationship with *C. africana* and the ribbing of the calyx, which I suspect is the cause of the statement, is of a quite different order of distinctness. I think *C. fischeri* is probably a form of *C. monoica*. It is most unlikely that there are any duplicates of the type extant and its identity will remain doubtful. Massaini has not been traced and is probably just a locative of Masai.

10. **C. ellenbeckii** *Gürke* in Engl., V.E. 1(1): 171, fig. 139 (1910) nomen* & apud F. Vaupel in E.J. 48: 526 (1912); E.P.A.: 767 (1961). Types: Ethiopia, 'Gallahochland', Boran, *Ellenbeck* 2057 & 2092 (B, syn.†)

Shrubs 1–5 m. tall with dark grey bark; young branches sparsely pubescent or villous but soon glabrous. Leaves often on very short shoots, almost round, obovate or oblong, 0.8–3 cm. long, 0.4–2.5 cm. wide, rounded to ± acute at the apex, narrowed at base into the petiole, margin crenate-undulate to irregularly dentate, often densely velvety when young, later both sides pubescent-tomentose, white beneath but with yellow brown hairs on the venation, not or only slightly rough at first but more distinctly so at maturity; petiole 1–5 mm. long. Flowers unisexual, rather fragrant, opening in the evening; ♂ flowers in 5–6-flowered inflorescences, sometimes appearing terminal but in forks of branches; ♀ flowers solitary. Male: Calyx very narrowly funnel-shaped, 11 mm. long with irregular short lobes 1–2 mm. long, with short spreading white hairs and pale fulvous tomentum outside. Corolla cream, glabrous; tube subcylindrical, 1.5 cm. long, ± widened above to 2.5 mm. at throat; lobes oblong, 5–6 mm. long, 3.5 mm. wide, entire or 2–3-fid (natural?); anthers just exserted from tube, the free parts of the filaments very short; rudimentary ovary ± 1 mm. long with no style. Female with calyx similar but wider, 9 mm. long, 5.5 mm. wide, with irregular teeth 1–2 mm. long; corolla-lobes 6 mm. long, 2 mm. wide. Anthers just exserted on very short filaments. Ovary ovoid, 2.5 mm. long. Styles ± 9 mm. long; stigmatic branches ± 5 mm. long, exserted. Fruit translucent orange, ± globose, 1.2 cm. diameter, obscurely ribbed, in a cup 1.6 cm. wide 6 mm. tall, subsessile. Fig. 4/5, p.16.

KENYA.   Northern Frontier Province: Dandu, 19 Mar. 1952, *Gillett* 12592! & 2 Apr. 1952, *Gillett* 12674! & 4 Apr. 1952, *Gillett* 12684! & Mandera, 30 km. from Ramu on Malka Mari [Murri] road, 8 May 1978, *Gilbert & Thulin* 1562!
DISTR. **K** 1; S. Ethiopia
HAB.   *Commiphora-Acacia* woodland and bushland, in limestone valley; 400–780 m.

NOTE.   Gürke stated the pistil was unknown because he assumed it was eaten by insects but he probably had male flowers. Warfa, Acta Univ. Upsal. 174: 31, 33 (1988) considers *C. ellenbeckii* is not distinct from *C. monoica*.

* After 1 Jan. 1908 names were no longer valid if based on figures alone.

11. C. sp. A

Shrub about 2 m. tall; young shoots pale ferruginous pubescent but soon glabrous, slender, blackish and with prominent nodular petiole-base remains; older stems ± smooth, bronze-brown, quite glabrous. Leaves alternate; blades narrowly elliptic, 1–5.5 cm. long, 0.7–2 mm. wide, rounded at the apex and minutely mucronulate, cuneate at the base, somewhat rough above, puberulous on the main nerves beneath; petiole 2.5–3.5 mm. long. Only poorly preserved detached ♀ flowers known. Calyx ± funnel-shaped, 9–10 mm. long including ± 2 mm. long triangular lobes, distinctly ribbed above in the dry state, pubescent with ± stiffish hairs outside, puberulous inside. Corolla-tube slender, ± 9 mm. long; lobes ?4 mm. long. Ovary obovoid, glabrous; style 5 mm. long, thickened at base; branches 2 mm. long; stigmas linear, 4 mm. long, ± flattened. Fruits solitary, sessile, terminal or pseudo-axillary, ovoid, 8 mm. long, 7 mm. wide (immature), rounded with short apiculum, sitting in a cupular calyx, at first ± 8 mm. tall, later 1.3 cm. wide; ± tomentose and pubescent outside, adpressed pubescent inside.

TANZANIA. Mpwapwa District: Godegode, 8 Feb. 1933, *B.D. Burtt* 4568! & near Gulwe, 8 Dec. 1925, *Peter* 45658!
DISTR. T5; not known elsewhere
HAB. *Commiphora-Cordyla* thicket on shallow stony alluvium, also on small outcrops; 850–900 m.

NOTE. Only two specimens of this have been seen, distinctive by its ?solitary sessile flowers and narrow leaves. It must be related to *C. monoica* Roxb. but it seems unlikely to be merely a variant of it. The true nature of the inflorescence will only be apparent when adequate flowering material is available; the Peter sheet has only a few imperfect detached flowers. It is probably genuinely uncommon since the area has been quite extensively collected by a number of botanists.

12. C. sp. B

Shrub 3–5 m. tall; youngest stems pale yellowish; innovations with some adpressed scurfy hairs but soon glabrous, grey-brown, ridged and lenticellate. Leaves rather thick, broadly elliptic to obovate, 1–6 cm. long, 0.6–4 cm. wide, broadly rounded to subacute at the apex, rounded to cuneate at the base, undulate to ± crenate-dentate or ± entire, glabrous. Fruiting inflorescence remnants 2–3 cm. long including peduncle, ± glabrous. Remnants of fruiting calyx 4 mm. long, 7 mm. wide, constricted at base, margin undulate-lobed, glabrous. Fruit orange.

KENYA. Lamu District: E. edge of Dodori Reserve, about 8 km. W. of Kiunga, 26 Oct. 1980, *Kuchar* 13691! & same area, about 6 km. W., 31 Oct. 1980, *Muchiri* 567!
DISTR. K 7
HAB. *Acacia-Combretum* woodland with almost continuous flat canopy, on sandy soil.

NOTE. Despite one having elliptic leaves and the other obovate I suspect the two specimens cited are the same but more material is needed to establish their identity. They may be forms of *C. crenata*.

13. C. crenata *Del.*, Fl. Egypte: 195 [51] (1813), t. 20 (1826); DC., Prodr. 9: 479 (1845); Bak. & Wright in F.T.A. 4(2): 17 (1905); Warfa, Acta Univ. Upsal. 174, III: 3 (1988) & in Nordic Journ. Bot. 8: 614 (1989). Type: Egypt, cultivated in Cairo gardens, *Delile* (MPU, holo.!)

Shrub or small tree 1.5–9 m. tall; young shoots with dense adpressed hairs and usually long ± spreading hairs as well, later ± glabrous or long hairs persistent on upper 15 cm. of stem; bark pale buff-brown or grey, often with ± prominent nodular petiole-bases; in subsp. *shinyangensis* young stems, young leaves, inflorescence-axes and buds ferruginous velvety. Leaf-blades broadly elliptic, obovate-oblong or almost round, less often oblong-lanceolate, 1–10.8 cm. long, 0.8–8 cm. wide, rounded, emarginate or subacute at apex, or rounded to cuneate at base, entire or frequently crenate to dentate, puberulous to pubescent above and beneath but not or only slightly asperous, or ± glabrous; petiole 0.5–1 cm. long with indumentum like stems when young. Inflorescences 2.5–3 cm. long, ± sessile to shortly pedunculate; peduncle and secondary axes 0.5–4 cm. long, ± velvety and with long hairs; pedicels 2–3 mm. long. Male: buds very shortly bluntly acuminate, adpressed pubescent, particularly at the apex; calyx-tube conic-campanulate 3.5–4.5 mm. long; lobes broadly triangular, 1.2–1.5 mm. long; corolla cream or greenish yellow; tube 3.5–4 mm. long; lobes 4–5(–6), ovate-oblong, 4–4.5(–6) mm. long, 1–2.5 mm. wide; filaments 1.5–6 mm. long, rudimentary ovary under 1 mm. long; style absent. Hermaphrodite: calyx 5.5–6.5 mm. long including 1–1.3 mm. long ovate lobes, ± glabrous outside, minutely pubescent inside; corolla-tube 4–5 mm. long; lobes 4–5(–6), oblong to narrowly ovate, 3–4 mm. long, 1.5–2 mm. wide; filaments 1.5–2 mm. long; style 2–3.5 mm. long; branches very short, up to 1.3 mm. long; stigmatic arms 7 mm. long. Fruit red, ovoid-globose, 0.7–1.3 cm. long, 6–8 mm. wide, apiculate; calyx cupular, 6.5–9 mm. wide, lobed, glabrous or pilose.

KEY TO INFRASPECIFIC VARIANTS

1. Leaf-blades narrowly elliptic to oblong-lanceolate, entire; young stems, young leaves, inflorescence-axes and buds ferruginous velvety . . . .   c. subsp. **shinyangensis**
   Leaf-blades more broadly elliptic, obovate-oblong or almost round, entire or toothed; young stems, etc. glabrous to pubescent and pilose but not densely ferruginous velvety 2
2. Leaves and young stem ± glabrous or with sparse short and/or long hairs . . . .   a. subsp. **crenata**
   Leaves glabrous to pubescent; young shoots mostly densely adpressed pubescent and usually ± long pilose as well   b. subsp. **meridionalis**

a. subsp. **crenata**; Warfa, Acta Univ. Upsal. 174, III: 5, figs 1, 2 (1988) & in Nordic Journ. Bot. 8: 614 (1989)

Leaves and young stems mostly ± glabrous or with sparse indumentum; blades broadly elliptic, obovate-oblong or almost round, entire or toothed.

UGANDA. W. Nile District: Nile Flats, Laropi, Dec. 1931, *Brasnett* 336!
DISTR. U1; Sudan, Ethiopia, Somalia, also Yemen, Oman, Iran and India
HAB. Riverine flats; 600 m.

SYN. [*C. myxa* sensu Forssk., Fl. Aegypt.-Arab.: LXIII (1775), *non* L.]
     C. zedambae Martelli, Fl. Bogos.: 58 (1886); Bak. & Wright in F.T.A. 4(2): 15 (1905); E.P.A.: 769 (1961). Type: Ethiopia, Eritrea, Bogos, Shotel, on the slopes of Zedamba (Tsad-Amba), *Beccari* 115 (FT, lecto., K, isolecto.!)
     C. lowriana Brandis, Indian Trees: 479 (1906). Type: India, Rajputana, Merwara, *Duthie* 4754 (K, lecto.!, BM, CAL, isolecto.)
     C. sp. 1 sensu I.T.U., ed. 2: 49 (1952)

b. subsp. **meridionalis** Warfa, Acta Univ. Upsal. 174, III: 7, figs 2, 3 (1988) & in Nordic Journ. Bot. 8: 617 (1989). Type. Kenya, Northern Frontier Province, South Horr, *J. Adamson* 24 (K, holo.!, EA, iso.!)

Leaves and young stems glabrous to pubescent and pilose but not densely ferruginous velvety; blades as in subsp. *crenata*. Fig. 4/6, 7, p.16.

UGANDA. Acholi District: SE. Imatong Mts., Agoro [Agora], 9 Apr. 1945, *Greenway & Hummel* 7330!; Karamoja District: Rupa, Apr. 1958, *J. Wilson* 408! & foot of Karamoja rift near Moroto R., *Brasnett* 86!
KENYA. Northern Frontier Province: Mathews Range, Olkanto, 13 Dec. 1944, *J. Adamson* 46 in *Bally* 4347!; Machakos District: Kibwezi, 16 June 1909, *Scheffler* 230!; Kitui/Kilifi Districts: Lali Hills, Galana R., 30 Mar. 1963, *Thairu* 110!
TANZANIA. Masai District: 12.8 km. S. of Namanga, 12 Dec. 1959, *Verdcourt* 2524! & Ngorongoro, above the Ol Balbal Depression, 25 Nov. 1956, *Tanner* 330!; Mbulu District: Yaida valley, Laangada Damari, 22 Jan. 1970, *Richards* 25257!
DISTR. U1; K1, 2, 4, 6, 7; T2, 5, 7, 8; Somalia
HAB. *Acacia-Commiphora-Salvadora* bushland on pale stony soil, secondary bushland along stony river beds, riverine woodland; 60–1500 m.

SYN. [*C. ovalis* sensu Bak. & Wright in F.T.A. 4(2): 16 (1905), quoad *Kassner* 560, *non* R.Br.]
     C. chisimajensis Chiov., Fl. Somala 2: 311, t. 81 (1932); E.P.A.: 767 (1961). Type: Somalia (S.), Kismayu (Chisimaio), *Senni* 534 (FT, holo.!)
     ?*C. bakeri* sensu Chiov., Racc. Bot. Miss. Consol. Kenya: 84 (1935), quoad *Mearns* 515, *non* Britten sensu stricto]
     C. sp. 2 sensu I.T.U., ed. 2: 49 (1952), quoad *Eggeling* 793
     [*C. crenata* sensu K.T.S.: 69 (1961); Vollesen in Opera Bot. 59: 75 (1980) et annot. mult., *non* Del. sensu stricto]
     C. sp. aff. crenata Del.; Hepper, Jaeger et al., Annot. Check-list Pl. Mt. Kulal: 86 (1981)

NOTE. Forms with very small leaves and very condensed ± sessile inflorescences exist, often also lacking the longer hairs, e.g. *Hepper & Jaeger* 6756 (Kenya, Northern Frontier Province, South Horr, 12 Nov. 1978). They are perhaps analogues of *C. quercifolia* Klotzsch. The species as a whole is very closely allied to *C. sinensis* and the stems ± identical.

c. subsp. **shinyangensis** Verdc. a subsp. *crenata* foliis angustioribus ellipticis usque oblongo-lanceolatis 2–6.5(–9) cm. longis 1–2.7(–3.7) cm. latis primo subtus dense ferrugineo-velutinis, ramis juvenilibus axibusque inflorescentiis etiam dense ferrugineo-velutinis differt. Typus: Tanzania, Shinyanga, *Koritschoner* 2092 (EA, holo.!, K, iso.!)

Young leaves and stems, also inflorescence-axes and buds ferruginous velvety; blades narrowly elliptic to oblong-lanceolate, entire.

TANZANIA. Shinyanga, *Koritschoner* 2092!
DISTR. T1; not known elsewhere
HAB. Presumably woodland

NOTE. I had originally treated this as a separate species but some material I cited is considered by Warfa to be within the variation of subsp. *meridionalis*. Only the specimen cited above is too diverse to be included.

14. **C. longipetiolata** *Warfa*, Acta Univ. Upsal. 174, fig. 4C & 174, IV: 1, fig. 1 (1988) & in Nordic Journ. Bot. 9: 251 (1989). Type: Kenya, Northern Frontier Province, Dadaab to Wajir road, 15 km. S. of Sabule, *Gillett & Gachathi* 20644 (K, holo.!, BR!, EA!, UPS, iso.)

Shrub to 3 m.; stems ± straw-coloured, ridged when dry; very young parts with long ± adpressed hairs and pale brownish tomentum of shorter hairs, later ± glabrous save for very scattered long hairs. Leaves dark green, drying blackish green; blades ovate to broadly elliptic, 1.5–7.5 cm. long, 1–6 cm. wide, subentire to irregularly coarsely dentate-crenate, smooth to the touch above but actually with minute conical tubercles and cystolith groups, with a few long white hairs on the midrib beneath; petiole drying blackish, (2.2–)3–5 cm. long. Cymes dense, terminal or axillary, 1–1.5 cm. long and wide, densely pubescent; peduncle 1–7 mm. long; pedicels 1–2 cm. long, tomentose. Buds obovate, 6.5 mm. long, hairy at apex. Male flowers: calyx 7–8 mm. long, narrowed near the base, 3–4-lobed; corolla cream; tube very narrowly funnel-shaped, 9 mm. long; lobes 5–6, ovate-oblong, 4–6 mm. long, 3.5 mm. wide; stamens 5–6, with filaments 2.5–3 mm. long and anthers 3 mm. long; rudimentary ovary and style under 2 mm. long. Female flowers similar. Fruit orange, ovoid, 1.5–1.8 cm. long, 1–1.1 cm. wide; calycine cup 1.1–1.2 cm. wide, obscurely toothed.

KENYA. Northern Frontier Province: Dadaab–Wajir road, 5 km. S. of Sabule, 1977, *Gillett* 21248! (♂) & 21249! (♀) & Dadaab to Wajir road, 15 km. S. of Sabule, 12 May 1974, *Gillett* 20644 (♂ & fruit)
DISTR. K1; not known elsewhere
HAB. *Acacia-Commiphora* mixed bushland on sand; 200 m.

15. **C. sinensis** *Lam.*, Tab. Encycl. 1: 423 (1792); Poir. in Encycl. Méth. Bot. 7: 49 (1806); I.M. Johnston in Journ. Arn. Arb. 32: 11 (1951); Heine in Adansonia, sér. 2, 8: 186 (1968); Meikle in Israel Journ. Bot. 20: 22 (1971); Täckh., Student's Fl. Egypt, ed. 2: 436, t. 152 (1974); Warfa, Acta Univ. Upsal. 174, V: 2, fig. 1, 2 (1988) & Nordic Journ. Bot. 9: 649, figs. 1,2 (1990); Martins in F.Z. 7(4): 69 (1990), pro parte. Type: India, *Sonnerat* (P-LAM, holo., microfiche!)

Tangled shrub or usually a spreading often several-stemmed tree 3–10(–12) m. tall; young stems ridged, slightly pubescent; young bark pale, smooth but later brownish grey or yellowish brown to almost black, densely longitudinally fissured and very rough; innovations often ± velvety, fawn-coloured. Leaf-blades typically narrowly oblong-elliptic or oblanceolate to oblanceolate-obovate, mostly rather longer than in species 16, 1.5–11 cm. long, 0.6–4 cm. wide, rounded at the apex, rounded to cuneate at the base, entire or crenate near apex or rarely crenate all round, glabrous to pubescent and sometimes slightly scabrid above, nearly always with longer hairs in the nerve-axils and along the midrib beneath; occasionally somewhat asperous on both surfaces; petiole distinctly developed, up to 1.3(–1.8) cm. long. Flowers sweet-scented in rather laxer panicles than in species 16, 6.5 × 7 cm.; peduncles and secondary axes up to 1.5–3 cm. long; pedicels 2 mm. long. Calyx glabrous to pubescent outside; tube cylindric-campanulate, 4–4.2 mm. long including 0.5–1 mm. long low rounded lobes. Corolla creamy white; tube 5 mm. long. Male: corolla-lobes oblong, 4.5 mm. long, 1.5 mm. wide; filaments exserted 3.5 mm. Female: corolla-lobes 3–4 mm. long, 1.3–1.5 mm. wide; filaments 2.2 mm. long; style 2.5 mm. long with branches 2 mm. long and stigmatic branches 2–4 mm. long. Fruit yellow, orange or bright red, 0.7–1.3 cm. long, 7–9 mm. wide, mucilaginous and edible. Fig. 4/8, 9, p.16.

UGANDA. Bunyoro District: E. Lake Albert, 30 Dec. 1906, *Bagshawe* 1402!
KENYA. Northern Frontier Province: Tana R., Masabubu, 1 Nov. 1945, *J. Adamson* 185!; Masai District: foothills of Nguruman Escarpment, Oloibortoto, 5 Aug. 1962, *Glover & Samuel* 3234!; Kilifi District: Jilore, Viragoni area, 20 Mar. 1973, *Sangai* in *E.A.H.* 15595!

TANZANIA. Masai District: Mto wa Mbu, 12 Feb. 1964. *Greenway* 11186!; Pare District: Gonja Maore, July 1955, *Semsei* 2108!; Mpanda District: Lake Rukwa, Zimba, 2 Nov. 1933, *Michelmore* 720!

DISTR. U2; K1, 2, 4, 6, 7; T1–8; Ghana, Togo, Nigeria, Chad, Sudan, Ethiopia, Somalia, Angola, Namibia, Israel, Jordan, Saudi Arabia, Egypt, Pakistan, India, Ceylon

HAB. Mainly riverine, in tree belt or tangled scrub, sometimes in scrub and open thorn-bush of *Acacia*, *Commiphora*, etc. subject to inundation, ground-water forest; 0–1800 m.

SYN. *C. reticulata* Roth in Roem. & Schultes, Syst. Veg. 4: 454 (1819); Roth, Nov. Pl. Sp. Ind. Or.: 124 (1821), *non* Vahl (1807), *nom. illegit.* Type: "India orientali", *Heyne* (ubi, holo., K-WALL 895-2, iso.!)

 C. rothii Roem. & Schultes, Syst. Veg. 4: 798 (1819); DC., Prodr. 9: 480 (1845); Wight, Ic. Pl. Ind. Or., t. 1379 (1848); C.B. Cl. in Hook.f., Fl. Brit. Ind. 4: 138 (1883); Boiss., Fl. Orient. 8: 350 (1888); Bak. & Wright in F.T.A. 4(2): 45 (1905); F.P.S. 3: 78, fig. 16 (1956); Heine in F.W.T.A., ed. 2, 2: 32 (1963); Riedl in Fl. Iran. 48/15: 6 (1967); Meikle in Israel Journ. Bot. 18: 142 (1969). Type as for last

 C. subopposita DC., Prodr. 9: 480 (1845); A.Rich., Tent. Fl. Abyss. 2: 81 (1850); Hiern, Cat. Afr. Pl. Welw. 1: 713 (1898), based on *Cornus sanguinea* sensu Forssk. *non* L., 'Arabia, Yemen, Lohaja, Surdud and Hadie'

 C. gharaf Aschers. in Sitz. Ges. Nat. Freunde Berlin 1879: 46 (1879) & in Verh. Bot. Ver. Brandenb. 21: 69 (1880); Engl., Hochgebirgsfl. Trop. Afr.: 351 (1892); Gürke in P.O.A. C: 335 (1895) & in E.J. 28: 461 (1900); Post, Fl. Pal. 2: 219 (1933); Schwartz in Mitt. Inst. Bot. Hamburg 10: 204 (1939); T.T.C.L.: 76 (1949); I.M. Johnston in Journ. Arn. Arb. 37: 292 (1956); Täckh., Students' Fl. Egypt: 157 (1956); K.T.S.: 69 (1961); E.P.A.: 767 (1961). Based on *Cornus gharaf* Forssk., *nom. invalid.* and *C. sanguinea* Forssk., *non* L.

NOTE. Some of the distribution may be due to cultivation.

16. **C. quercifolia** *Klotzsch* in Peters, Reise Mossamb., Bot.: 247, t. 43 (1861). Type: Mozambique, Tete, *Peters* (B, holo.†)

Spreading shrub to small bushy tree 0.3–4.5(–6) m. tall, sometimes even said to be ± creeping; young shoots ridged, often pubescent, lenticellate, the innovations pale ferruginous pubescent; bark dark, grey or almost black, thick, corky, deeply longitudinally fissured; slash soft grey-brown outer area and soft yellow-brown inner. Leaves nearly always opposite, rarely alternate on some shoots; blades oblong or narrowly oblong to oblong-elliptic, 0.9–7.5(–10) cm. long, 0.7–4(–5) cm. wide, rounded to slightly indented at the apex and ± mucronulate, rounded to cuneate at the base, entire or very slightly crenate towards the apex, slightly rough above with short white hairs and with longer white hairs beneath, often quite dense or with longer hairs on both surfaces; venation ± reticulate beneath; mostly subsessile or petiole up to 7 mm. long. Flowers in ± sessile cymes or peduncle 0.8–1.5 cm. long; pedicels 0–1 mm. long becoming 4–5 mm. in fruit. Calyx densely adpressed pubescent or velvety, becoming glabrescent in fruiting stage; tube 3.5–4(–6.5) mm. long, shortly lobed. Corolla white. Male: corolla-lobes up to 6 mm. long; stamens exserted 3–5 mm.; style ± obsolete or if present then stigma not reaching the anthers. Female: corolla-tube 3.5–5 mm. long; lobes 2–4.5 mm. long, 1–1.5 mm. wide; filaments exserted 1.5 mm.; style 3 mm. long with branches 0.5–1.3 mm. long and stigmas 2–2.5 mm. long, well exserted. Fruit golden yellow or bright orange, ovoid with sharp beak-like style-base, 1–1.7 cm. long, 0.7–1 cm. wide, mucilaginous and edible; endocarps yellowish, ellipsoid, 6 mm. long, 4 mm. wide; fruiting calyx cupular, 9 mm. diameter. Fig. 4/10, 11, p. 16.

UGANDA. Karamoja District: Lotizan, Sept. 1943, *Dale* U365! & Kanamugit, Feb. 1936, *Eggeling* 2927! & base of Mt. Kadam [Debasien], Kokumongole [Kakumongole], Jan. 1936, *Eggeling* 2602!

KENYA. Northern Frontier Province: Dandu, 5 Apr. 1952 (♂), *Gillett* 12706! & 18 Mar. 1952 (♂), *Gillett* 12573!; Masai District: 24 km. on Ol Tukai–Namanga road, 14 Dec. 1959, *Verdcourt* 2568!; Kwale District: near Taru, 6 Sept. 1953, *Drummond & Hemsley* 4185!

TANZANIA. Masai District: 13 km. S. of Namanga, 11 Dec. 1959, *Verdcourt* 2516!; Lushoto District: Mkomazi, 23 Apr. 1934, *Greenway* 3963!; Iringa District: 155 km. on Iringa–Dodoma road, 26 Feb. 1961, *Verdcourt* 3074!

DISTR. U1; K1, 4, 6, 7; T1–3, 5–8; Mali, Niger, Mauritania, Chad, Ethiopia (Ogaden), Mozambique, Zimbabwe, Arabia, India and Ceylon

HAB. Grassland with scattered trees, *Acacia-Commiphora*, etc. bushland and thicket, lava desert, on black and red soils, never riverine; 60–1500 m.

SYN. [*C. perrottetii* sensu Wight, Ic. Pl. Ind. Or., t. 1381 (1848); C.B. Cl. in Fl. Brit. Ind. 4: 138 (1883), *non* A.DC.]

 [*C. rothii* sensu Bak. & Wright in F.T.A. 4(2): 19 (1905), pro parte, *non* Roem. & Schultes sensu stricto.]

 C. nevillii Alston in Handb. Fl. Ceylon 6 (Suppl.): 199 (1931); Warfa, Acta Univ. Upsal. 174, v: 7, fig. 3, 4 (1988) & in Nordic Journ. Bot. 9: 652 (1990). Type: Ceylon, Kuchaveli, *Nevill*, distributed as *Alston* 578 (PDA, holo., K, iso.!)

[*C. gharaf* sensu I.T.U., ed. 2: 48 (1952), quoad *Dale* 365, *Eggeling* 1707 et auctt. et adnot. mult., *non* Aschers.]

C. sp. 3 sensu I.T.U., ed. 2: 49 (1952)

C. sp. 4 sensu I.T.U., ed. 2: 50 (1952)

C. sp. near quarensis Gürke sensu K.T.S.: 71 (1961)

[*C. crenata* sensu Vollesen in Opera Bot. 59: 75 (1980), *non* Del.]

[*C. sinensis* auctt. et adnot. mult., e.g. Martins in F.Z. 7(4): 69 (1990) pro parte *non* Lam.]

NOTE. I had at first considered this to be a well marked variety or ecological subspecies of *C. sinensis*. The two are easily told apart, however, and there are but few intermediates. Unfortunately the earliest name is from the Zambesiaca area where the distinction is least marked but an original suggestion that *C. nevillii* should be retained as the name is untenable. In the Flora Zambesiaca area Martins has rather understandably adhered to a single variable *C. sinensis* concept.

17. **C. somaliensis** *Bak.* in K.B. 1894: 28 (1894); Bak. & Wright in F.T.A. 4(2): 18 (1905); Chiov., Result. Sci. Miss. Stef.-Paoli, Coll. Bot.: 119, 215 (1916) & in Agric. Colon. 20: 46 (1926) & in Fl. Somala 1: 225 (1929) & 2: 308 (1932); K.T.S.: 71 (1961); E.P.A.: 769 (1961); Warfa, Acta Univ. Upsal. 174: 45, figs. 5/c, 9, 10 (1988). Type: Somalia, coast near 'Kisinga' (Kismayu fide original label and Cufodontis), *Kirk* (K, holo.!)

Scrambling shrub or bushy tree with several stems from the base, (0.3–)1.8–6 m. tall; bark pale brown, becoming grey and fissured; young stems ridged, grey, densely pubescent with adpressed white hairs and longer spreading ones, later glabrescent. Leaf-blades pale green in life, obovate-oblong, 1.7–7.5(–10) cm. long, 1–5.3 cm. wide, rounded to acute at the apex, ± cuneate at the base, entire to shallowly crenate towards apex, ± fleshy, obscurely scabrid above with cystoliths but no rough hairs, glabrous or pubescent, adpressed pubescent beneath or almost glabrous; petiole 6–11(–15) mm. long, ± pilose. Flowers sweet-scented, opening in the evening, sessile in condensed cymes arranged in small inflorescences ± 6 cm. wide; peduncles ± 3 cm. long, with secondary axes ± 1.5 cm. long. Male: calyx campanulate; tube 3–6 mm. long, constricted near the base when dry, sparsely pubescent outside, shortly adpressed pilose inside with 2–3 semicircular lobes ± 1 mm. long; corolla cream or yellowish, glabrous outside; tube cylindrical, 6 mm. long; lobes oblong-obovate, 6 mm. long, 2.5 mm. wide, narrowed below; stamens exserted; filaments 2.5–3.5 mm. long with adpressed part pilose; rudimentary ovary depressed globose, 1 mm. tall, 1.5 mm. wide; style obsolete. Female: calyx campanulate; tube 4 mm. long; lobes 1 mm. long; corolla-tube fusiform, 5 mm. long; lobes narrowly obovate-oblong, 3.5 mm. long, 1.3 mm. wide; filaments 1 mm. long with adnate part shortly pilose; anthers present but sterile; ovary fusiform, 4 mm. long, 1.5 mm. wide; style stout, 1–2 mm. long, with branches 1.5–2.5 mm. long, the 4 stigmatic branches ± 3.2 mm. long. Fruit orange, ovoid-conic, 1–1.7 cm. long including the long-drawn-out acute beak, sitting in the 7–11 mm. wide, venose accrescent calyx. Fig. 5/1, 2, p. 24.

KENYA. Northern Frontier Province: 13 km. N. of Garissa on Saka track, 3 Mar. 1973, *Sangai* 940!; Kilifi District: Vipingo, 16 Dec. 1953, *Verdcourt* 1062!; Lamu District: SE. side of Manda I., Takwa, 3 Oct. 1957, *Greenway* 9271! & Lamu Town, E. side, 14 Feb. 1956, *Greenway & Rawlins* 8904!

DISTR. **K** 1, 7; Somalia (S.)

HAB. Sandy open areas near high-tide mark with *Guettarda*, etc., dunes, thicket on coral rag and other coastal bushland, also open *Acacia* scrub on red sandy soil; 0–15(–150) m.

NOTE. Fruit edible. This species is closely related to *C. quercifolia* Klotzsch but differs in indumentum, habit and ecology.

18. C. sp. C

Much-branched shrub; young branches compressed, densely covered with long very pale brownish hairs; previous season's growth soon glabrous and with grey-brown bark; petiole-bases remaining and with strong angular decurrent ridges from them so that in places the stem is 3-angled. Leaf-blades rounded ovate, ± 12.5 cm. long, 10 cm. wide, acute at the apex, sometimes unequal at the base but ± rounded, glabrous save for some sparse long hairs; petiole 7 cm. long, hairy as in the young stems. Inflorescences terminal, lax, much branched, 9 cm. long, 8 cm. wide, the axes densely hairy as in young stems; secondary peduncles, etc. up to 3 cm. long; pedicels 0–2 mm. long. Buds obovoid, 5–6 mm. long, glabrous save apex. Male flowers: calyx narrowly funnel-shaped, constricted at the extreme base, silky pilose inside, 7 mm. long including ± 3 mm. long and 2 mm. wide rounded lobes; corolla-tube narrowly cylindrical, 5 mm. long, with a few hairs in the throat; lobes 5, oblong, 4–4.5 mm. long, 1.6 mm. wide, obtuse; filaments ± 3 mm. long; anthers 1.2 mm. long; ovary rudimentary, globose, 1 mm. diameter; style absent. Female flowers and fruit unknown.

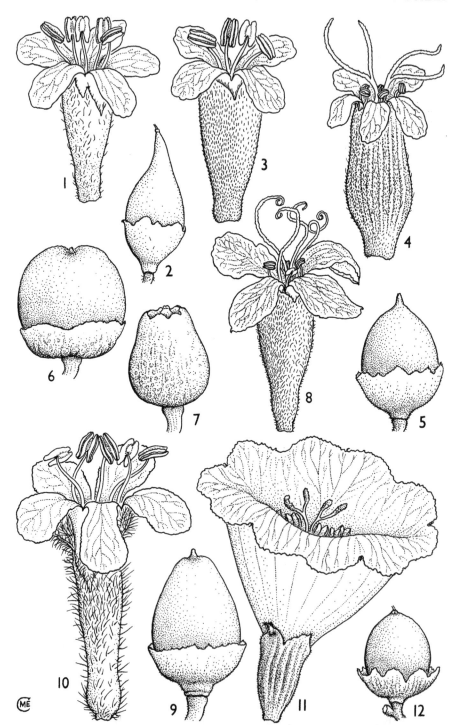

FIG. 5. *CORDIA SOMALIENSIS*—**1**, ♂ flower × 4; **2**, fruit, × 2. *C. GUINEENSIS* subsp. *GUINEENSIS*—**3**, ♂ flower, × 4; **4**, ♀ flower, × 4; **5**, fruit, × 2. *C. BALANOCARPA*—**6**, fruit, × 2. *C. FAULKNERAE*—**7**, fruit, × 2. *C. TORREI*—**8**, ♀ flower, × 4; **9**, fruit, × 2. *C. CHAETODONTA*—**10**, ♂ flower × 4. *C. AFRICANA*—**11**, flower, × 3; **12**, fruit, × 2. 1, from *Verdcourt* 1062; 2, from *Faden* 74/1088; 3, from *Irvine* 2679; 4, from *Irvine* 4826; 5, from *Morton* in *G.C.* 9259; 6, from *Burtt* 5654; 7, from *Verdcourt* 5299; 8, from *Torre & Paiva* 9261; 9, from *Peter* 59879; 10, from *Cribb, Grey-Wilson & Mwasumbi* 11202; 11, from *Meyerhoff* 3-0; 12, from *Kingua* 12. Drawn by Mrs Maureen Church.

TANZANIA. Lindi District: Mingoyo, 23 Mar. 1943, *Gillman* 1365!
DISTR. **T** 8; known only from the above gathering
HAB. Thicket on orange sands; near sea-level
NOTE. Appears to be an undescribed species.

19. **C. sp. D**

Deciduous shrub with slender ridged pale purplish brown stems, glabrous save for youngest innovations which are ferruginous-tomentose; short lateral shoots with evident blackish petiole scars. Mature leaves unknown, specimen flowering whilst leaves very young; immature leaf-blades narrowly oblong-elliptic, 1.6 cm. long, 5.5 mm. wide, acuminate at the apex, cuneate at the base, ferruginous pubescent on midnerve above and nervation beneath; petiole short. Inflorescences 2–3-fid, 2–3 cm. long and wide, terminating lateral branches; peduncles and secondary branches 0.5–1.5 cm. long; pedicels 0.5–1 mm. long; all axes ferruginous pubescent. Male: buds obovoid, pubescent at apex; calyx campanulate; tube ± 2.2 mm. long; lobes broadly triangular, 1.3 mm. long, slightly pubescent outside, puberulous inside; corolla pale yellow; tube 2.2 mm. long; lobes much longer, oblong to oblong-lanceolate, 3.5–4 mm. long, 2 mm. wide, reflexed; filaments 1.5 mm. long; anthers 2 mm. long; rudimentary ovary cushion-like; style absent. Female flowers and fruit unknown.

TANZANIA. Masasi, 12 Dec. 1942, *Gillman* 1244!
HAB. *Brachystegia* woodland on grey sandy loam; ?500 m.

20. **C. guineensis** *Thonn.* in Schumach. & Thonn., Beskr. Guin. Pl.: 128 (1827); DC., Prodr. 9: 480 (1845); Bak. & Wright in F.T.A. 4(2): 17 (1905); Heine in F.W.T.A., ed. 2, 2: 320 (1963). Type: Ghana [Guinea] coast, *Thonning* (G-DC, isosyn.)

Shrub or small tree to 3 m., or sometimes a liane; branches densely pubescent with short adpressed or ± spreading greyish hairs, later glabrous, usually with distinctly nodular petiole-bases. Leaf-blades oblong, obovate, elliptic or ± round, 1.5–9.5 cm. long, 1–8 cm. wide, shortly acuminate or rounded with short mucro at the apex, rounded, truncate or sometimes subcordate at the base, discolorous, subcoriaceous, entire or obscurely crenate to remotely dentate, particularly near the apex or with short subulate emergences (?hydathodes), with scattered white setae above and more densely setulose-pubescent beneath or in W. Africa sometimes densely pubescent on both surfaces, sometimes ± shiny above with venation impressed; petiole mostly 1–2.5 cm. long. Flowers opening at night, scented, in rather short dense pubescent cymes up to 9 cm. wide; peduncles 2–4.5(–7) cm. long; secondary axes ± 1(–3) cm. long; pedicels very short or obsolete, Buds with calyx strongly constricted at the apex, sometimes with recurved subulate lobes at tip. Calyx-tube tubular, becoming cupular in fruit, 4–6.5 mm. long, obscurely ribbed, grey-pubescent outside, silky inside, with ovate or deltoid lobes ± 0.8–2 mm. long, the true setaceous lobes having ± fallen. Corolla creamy yellow; tube urceolate-tubular or cylindrical to narrowly funnel-shaped, 4–7 mm. long; lobes oblong, 3–4 mm. long, 2–2.5 mm. wide, obtuse, ± spreading. Hermaphrodite: stamens not much exserted from the tube; ovary glabrous, tapering into the 4 mm. long style; style branches 2 mm. long; stigmas 7.5 mm. long. Male: filaments 3.5 mm. long, reaching tips of corolla-lobes, the anthers extending beyond; ovary rudimentary and style lacking. Fruit ovoid, 1.3–1.5 cm. long, tapering into indurated style remnant. Fig. 5/3–5.

SYN. *C. johnsonii* Bak. in F.T.A. 4(2): 13 (1905). Type: Ghana, Afram, *Johnson* 705 (K, holo.!)
     C. warneckei Bak. & Wright in F.T.A. 4(2): 13 (1905). Types: Togo, near Lome, *Warnecke* 100 & 308 (both K, syn.!, EA, isosyn.!)

NOTE. Subsp. *guineensis* occurs from Ivory Coast to Dahomey. It was used for fig. 5/3–5 in the absence of adequate material of subsp. *mutica*.

subsp. **mutica** *Verdc.* subsp. nov. a subsp. *guineensi* alabastris haud apiculis subulatis recurvis terminatis. Type: Tanzania, Uzaramo District, Mjimwema, *Vaughan* 2730 (BM, holo.!, EA, iso.)

Buds only slightly acuminate lacking distinct recurved subulate apiculae.

KENYA. Kwale District: Kaya Kinondo, 16 July 1987, *Robertson & Luke* 4907!; Lamu District: Boni Forest, about 20 km. W. of Kiunga, 17 Oct. 1980, *Muchiri* 452! (possibly — sterile)
TANZANIA. Uzaramo District: Mjimwema, 14 Jan. 1939, *Vaughan* 2730!; "Prov. Zanguebar", *Kirk*! (probably the island) & Zanzibar I., Kombeni Caves, July 1930, *Vaughan* 1413! & Dec. 1930, *Vaughan* 1732!
DISTR. **K** 7; **T** 6; **Z**; not known elsewhere

HAB.   Scarcely recorded (see note) but sometimes on limestone outcrops or at cave-edges; 0–10(–?50) m.

SYN.   [*C. dioica* sensu Bak. & Wright in F.T.A. 4(2): 16 (1905), quoad spec. *Kirk, non* DC.]

NOTE.   *Robertson & Luke* 5645 (Kenya, Lamu District, Boni Forest Reserve, 7 km. E. of Dodori R. at Mangai) has numerous well-developed buds which demonstrate that the slender apiculae terminating the calyx-lobes in West African material are not present. As Taton points out (F.C.B., Borag.: 14 (1971)) Heine (F.W.T.A., ed. 2, 2: 320 (1963)) was incorrect to include *C. kabarensis* De Wild. under *C. guineensis* — see synonymy of *C. monoica* on page 17.

The occurrence of this species on Zanzibar had been thought to be due to introduction despite the early date of the first collected specimen (1869). Its discovery in Tanzania and on the Kenya coast and recently of floral differences show it to be indigenous and such a distribution is not unprecedented. *Robertson & Luke* 4907 has very slender pale brown to purplish brown stems possibly climbing with the aid of the inflorescences, but despite its different facies I have no doubt as to its identity. Baker and Wright's '*C. dioica*' is a mixture of three species, one of which, the type, they did not see.

21.   **C. balanocarpa** *Brenan* in K.B. 4: 91, fig. on p. 92 (1949); T.T.C.L.: 75 (1949). Type: Tanzania, Dodoma District, Kazikazi, *B.D. Burtt* 5039 (K, holo.!, BR!, EA!, FHO, K!, iso.)

Deciduous scandent shrub to 5 m. with rigid branches covered with grey bark and numerous spur-like woody branchlets; young parts ± densely pubescent but eventually glabrous. Leaf-blades elliptic to oblong, obovate-elliptic or subobovate-oblong, 3–10 cm. long, 1.3–3.8 cm. wide, slightly emarginate or rounded at the apex, cuneate to rounded at the base, shining and smooth or scarcely scabrid above, densely pubescent or tomentellous beneath when young, eventually pubescent only on the prominent venation, rigid and ± coriaceous when adult, the margins very revolute when young; petiole short, 3–6 mm. long, sulcate above, ± pubescent. Flowers unisexual or ♂ in few-flowered congested inflorescences, terminal or at the ends of short lateral branches; pedicels 1–1.5 mm. long, tomentellous. Calyx obconic becoming ovoid-urceolate, 5.5–9 mm. long, 4–5.5 mm. wide near base but contracted to ± 3 mm. at mouth, not grooved but densely pubescent outside; lobes irregular, mostly broadly triangular, rounded or erose, 1–2 mm. long. Corolla pale yellowish; tube subcylindric, 8 mm. long, 1.5 mm. diameter above, glabrous outside, pilose inside above; lobes 4–5, obovate-oblong or subrhomboid, 7 mm. long, 3 mm. wide, obtuse at the apex. Stamens inserted in the throat; filaments 2.5 mm. long; anthers exserted. Ovary broadly ovoid-conic, ± 5.5 mm. tall, 4 mm. wide, glabrous; style filiform, 5 mm. long, bifurcate at apex, each branch 2.5 mm. long, again bifid, the branches all filiform, ± 2 mm. long. Fruits soft and jelly-like when ripe but hardening with age, broadly ovoid or subglobose, 1.2–1.6 cm. long, 1.1–1.6 cm. wide, glabrous, often with persistent style-base, sitting in the very accrescent rugulose cupuliform calyx which is woody, 1.2–1.5 cm. diameter, 8–9 mm. tall, puberulous outside, erose or scarcely lobed. Endocarp with 1–2 seeds. Figs. 5/6, p. 24; 6.

TANZANIA.   Shinyanga District: Samuye, Mar. 1936, *B.D. Burtt* 5654!; Dodoma District: Manyoni, 10 Apr. 1964, *Greenway & Polhill* 11494! & Kazikazi, 29 Oct. 1933, *B.D. Burtt* 5039!
DISTR.   **T** 1, 4, 5; not known elsewhere
HAB.   Deciduous thickets of *Commiphora-Grewia-Dalbergia-Combretum, Boscia, Pseudoprosopis* etc. on grey hard-pan soils, mbugas, grassland with *Commiphora, Lannea*, etc. bushland; 1140–1320 m.

NOTE.   The fruits are eaten by birds and humans. A characteristic element of the great thicket areas. More information is needed about the sexuality of the flowers — buds from *Shabani* 1154 (Tabora, 3 Dec. 1976) had sterile anthers —corolla-tube 3 mm. long and lobes oblanceolate 6 × 2.2 mm.; filaments 1.7 mm. long.

22.   **C. faulknerae** *Verdc.* in Bol. Soc. Brot., sér. 2, 53, fig. 1: 104 (1980); Vollesen in Opera Bot. 59: 75 (1980); Warfa, Acta Univ. Upsal. 174: 49, fig. 41D (1988). Type: Tanzania, Tanga District, Sawa, *Faulkner* 1567 (K, holo.!)

Climbing shrub to 9(–12) m. with branches of up to 20 cm. diameter; bark pale grey or yellowish. Leaves borne on young elongated branches or on very short lateral branches; blades oblong-elliptic or obovate-oblong, 2.5–10(–12.5) cm. long, 1.5–5.8(–7.3) cm. wide, rounded to subacute at the apex, cuneate to rounded at the base, entire, glabrous, eventually chartaceous, with distinct margin; petiole 1–2.5 cm. long, blackish on drying. Flowers scented, 4–5-merous, soon falling, borne in cymose-fasciculate inflorescences, each fascicule element ± 20-flowered; rachis 2–3(–10) mm. long; secondary axes 0–3(–8) mm. long; pedicels 2–5 mm. long, narrowly constricted at the apex. Corolla white, glabrous. Male: calyx conical, 5–6 mm. long, 1–2 mm. wide; corolla-tube 5.5–6.5 mm. long;

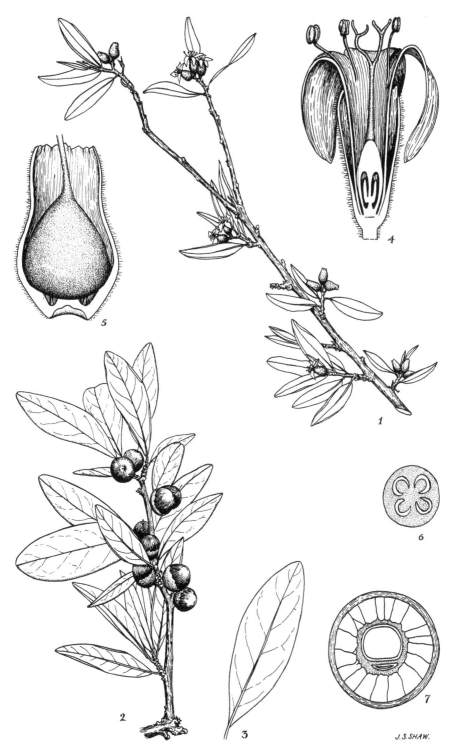

FIG. 6. *CORDIA BALANOCARPA*—**1**, flowering branchlet, × ½; **2**, fruiting branchlet, × ½; **3**, single leaf to show variation, × ½; **4**, flower in vertical section, × 5; **5**, young fruit with calyx partly removed, × 5; **6**, ovary in transverse section, × 10; **7**, mature fruit in transverse section (mesocarp apparently much shrunken and distorted). Drawn by J.S. Shaw.

lobes narrowly oblong-oblanceolate, 5 mm. long, up to 2 mm. wide at apex, retuse; free parts of filaments 2–3 mm. long; rudimentary ovary globose, 0.7 mm. diameter. Female: calyx narrowly obconic, 8–9 mm. long, 2.5 mm. wide, pubescent inside, slightly grooved with rounded lobes 1–1.5 mm. long, 0.5–2 mm. wide; corolla-tube 7.5 mm. long; lobes oblong-elliptic, 6.5 mm. long, 2 mm. wide, obtuse; antherodes 1 mm. long, the free part of the filaments 1.5 mm. long; ovary narrowly conical, 2.2 mm. long, the stipe 1 mm. long; style 6.5 mm. long, the main branches 0.5–1 mm. long and secondary branches 2 mm. long including the 2 mm. long stigmas. Immature fruits in subsessile clusters 4 cm. diameter, entirely included in the calyx; mature resembling oak acorns, ellipsoid, 1.8 cm. long, 1.4 cm. wide, mucronate at the apex, the cupuliform calyx 1 cm. long, 1.4 cm. wide; fruiting pedicels 3–4 mm. long, becoming woody. Seed ellipsoid, 1.4 cm. long, 9 mm. wide, reticulate-lamellate when dry. Figs. 5/7, p. 24; 7.

KENYA.   Teita District: Tsavo National Park East, Sala, 12 Nov. 1969, *Hucks* 1182!; Kwale District: between Kinango and Mariakani, R. Matumbi, 12 Apr. 1978, *Verdcourt* 5299!; Lamu District: Witu area, Feb. 1957, *Rawlins* 351!
TANZANIA.   Tanga District: Sawa, 11 Apr. 1956, *Faulkner* 1847! & 28 Sept. 1956, *Faulkner* 1921! & Machui, 7 Feb. 1957, *Faulkner* 1953!; Kilwa District: Kingupira, Lungonya R. Plain, 25 Mar. 1975, *Vollesen* in MRC 1944
DISTR.   **K** 7; **T** 3, 6, 8; Somalia (S.)
HAB.   Riverine bushland, thicket and forest, often with *Acacia elatior, Hyphaene, Rinorea, Mimusops, Spirostachys, Lecaniodiscus, Populus ilicifolia* etc., also foreshore and coastal wooded savanna; also in termite mound thicket; 0–125 m.

23. **C. torrei** *Martins* in Garcia de Orta, sér. Bot. 9: 71, t. 2 (1988) & in F.Z. 7(4): 68 (1990). Type: Mozambique, Monapo, 7 km. on new track from Namiolo towards Meserepane, *Torre & Paiva* 9261 (LISC, holo., K, iso.!)

Shrub ± 5 m. tall or tree to 18 m.; trunk 1 m. thick *fide* Peter; twigs grey, striate, lenticellate, glabrous, with very distinct nodular petiole-bases resembling short blunt spines. Leaf-blades elliptic to obovate, 6–12 cm. long, 3–9 cm. wide, acute to ± obtuse at the apex, ± cuneate at the base, entire to crinkly or coarsely dentate towards apex (much apparent crenation is due to insect damage), 3-nerved from the base, discolorous, at first matted velvety above but soon glabrous, velvety tomentose with matted grey hairs beneath; petiole (1.5–)2–4 cm. long, pubescent, channelled above. Flowers subsessile in terminal panicles of cymes 3–9 cm. long, 4–8 cm. wide. Female: calyx narrowly funnel-shaped, 7–9 mm. long, slightly sulcate, shortly tomentose outside, puberulous inside with 4–5 papery teeth; corolla probably white, glabrous; tube cylindrical, 6.5–8 mm. long; lobes 4, oblanceolate, 6–6.5 mm. long 2.5–3 mm. wide, obtuse; staminodes 4 with filaments 1 mm. long and antherodes 0.7 mm.; ovary ovoid, 2 mm. long; style 1–1.1 cm. long including 3–4 mm. long arms; stigmas linear, 4–6 mm. long. Fruiting inflorescences 5 cm. long; peduncle ± 4 cm.; pedicels thick, 3 mm. long. Fruit oblong-ovoid, ± 1.5 cm. long, 1 cm. wide, obtuse and excavated but mucronulate, glabrous, sitting in a calycine cup 6–8 mm. long, 1.3 cm. wide, contracted at junction with pedicel. Fig. 5/8, 9, p. 24.

KENYA.   Kwale District: Dzombo Hill, 9 Feb. 1989, *Robertson, Luke et al. MDE* 302!; Kilifi District: Kaloleni–Kilifi road, near Kaloleni, 1 Sept. 1959, *Verdcourt* 2414A! & Cha Simba, 14 Aug. 1989, *Luke & Robertson* 1876!
TANZANIA.   Lushoto District: Ngonja Mts, near Bamba, 22 June 1917, *Peter* 59879!
DISTR.   **K** 7; **T** 3; Mozambique
HAB.   Moist semi-deciduous forest with *Scorodophloeus, Cola, Combretum schumannii* etc.; also on rock-outcrop with *Casearia, Cola, Nesogordonia, Rinorea, Pandanus* and *Euphorbia*; 200–420 m.
NOTE.   Only a few specimens of this distinctive species have been seen from the Flora area. Practically all this type of lowland forest has been destroyed. Possibly related to *C. platythyrsa* Bak. from W. Africa, Zaire etc. but leaf-indumentum and fruits different.

24. **C. trichocladophylla** *Verdc.* sp. nov., ob folia discoloria et indumentum simile probabiliter affinis *C. torrei* Martins sed habitu frutescente inflorescentiis foliisque praecocibus surculis lateralibus brevibus insidentibus cortice caulium pallido suberoso differt. Typus: Tanzania, Lindi, Mlinguru, *Schlieben* 5761 (HBG, holo.! (♀), BR, iso.! (♂))

Shrub 1–2 m. tall; stems with pale lenticellate corky bark, obscurely striate and minutely transversely wrinkled in places, covered with short persistent nodular petiole-bases subtending short lateral leafy and flowering shoots or inflorescences with leaves still in very juvenile state; young shoots and inflorescence-axes completely covered with very

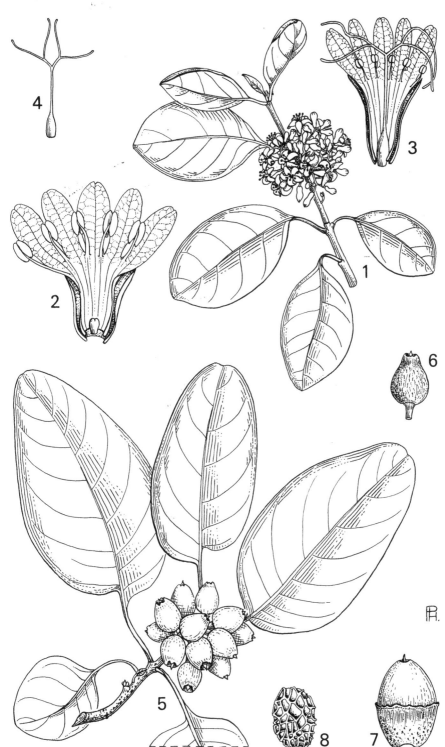

FIG. 7. *CORDIA FAULKNERAE*—**1**, flowering branch, × ⅔; **2**, ♂ flower, opened out, × 3; **3**, ♀ flower, opened out, × 3; **4**, gynoecium, × 2; **5**, fruiting branch, × ⅔; **6**, young fruit, × 1; **7**, mature fruit, × 1; **8**, seed, × 1. 1, 2, from *Hucks* 1182; 3, 4, from *Faulkner* 1847; 5, 6, from *Verdcourt* 5299; 7, 8, from *Faulkner* 1953. Drawn by Pat Halliday.

closely adpressed buff hairs. Leaf-blades oblong-elliptic, probably none fully developed in specimen seen, 2–5.5 cm. long, 1–2.5 cm. wide, acute to acuminate at the apex, cuneate at the base, at length very discolorous when dry, at first densely covered above with adpressed buff hairs but later with only a scattering and eventually probably glabrous, similarly but persistently hairy beneath and also with raised reticulate venation; petioles 0.5–2 cm. long, similarly buff-hairy. Cymes ± 20-flowered, ± 5 cm. long; peduncles and secondary axes 0.5–2 cm. long; pedicels 0–1.5 mm. long; buds narrowly conic. Male flowers: calyx-tube narrowly obconical, 5.5 mm. long; lobes triangular, 1.5 mm. long. Corolla-tube 5.5 mm. long; lobes elliptic, 6 mm. long, 3.2 mm. wide; filaments 4 mm. long; rudimentary ovary 0.6 mm. long; style absent. Female flowers: calyx adpressed buff hairy; tube obconical, slightly constricted about ⅕ from the base in dry state, feebly striate above, ± 6 mm. long, 3.5 mm. wide at apex; lobes triangular-ovate, 1–1.5 mm. long: corolla yellowish; tube 6–7 mm. long; lobes ovate-oblong, 4–5 mm. long, 3–3.5 mm. wide; filaments 0.2 mm. long; sterile anthers 1.1 mm. long; ovary 1.5 mm. long; style 4.5 mm. long with branches 2.5 mm. long and stigmas 8 mm. long. Fruits not known.

TANZANIA. Lindi District: Mlinguru, 21 Dec. 1936, *Schlieben* 5761!
DISTR. **T** 8; not known elsewhere
HAB. Light woodland; 60 m.

NOTE. Without additional material the affinities of this plant will remain obscure; it could just possibly, despite the very different habit, be only a variant of *C. torrei* Martins.

25. **C. chaetodonta** *Melchior* in N.B.G.B. 11: 676 (1932); T.T.C.L.: 76 (1949). Type: Tanzania, Ulanga District, Mahenge, Issongo, *Schlieben* 1718 (B, holo.†, BM, BR, HBG, iso.!)

Climbing shrub hanging over rocks, with slender grey-brown stems to several m. long; all parts of the plant covered with long ferruginous stiff hairs. Leaves alternate; blades elliptic, 4–15 cm. long, 3–8 cm. wide, acute or subobtuse at the apex, the tip shortly acuminate, ± rounded at the base, the margins repand or double-repand, the teeth ending in ± linear setae 0.5–2.5 mm. long, discolorous, drying dark above and laxly pubescent with adpressed hairs, brownish green or glaucescent beneath and setulose on the nerves; venation reticulate beneath; petiole 0.5–2.5 cm. long. Cymes terminal, dense, subglobose, ± 3.5 cm. diameter, spreading setulose; peduncle 2.5–4 cm. long; pedicels very short. Buds with 5 apical tails. Male flowers: calyx-tube narrowly funnel-shaped, 1.2–1.3 cm. long, not sulcate, the 4–5 lobes short, spreading ferruginous setose outside, adpressed white pilose inside, ending in setiform appendages ± 4 mm. long; corolla yellow; tube cylindrical, 1–1.8 cm. long, glabrous inside, setulose inside; lobes obovate-oblong, 5 mm. long, 2.5 mm. wide, with very sparse setae outside; stamens with free part of filaments 5 mm. long, exserted almost as long as corolla-lobes, with some spreading hairs at base, adnate parts with dense recurved hairs; rudimentary ovary 0.5 mm. diameter, glabrous, lacking a style. Female flowers and fruit unknown. Fig. 5/10, p. 24.

TANZANIA. Ulanga District: Kwiro, on slope below cathedral, 25 Jan. 1979, *Cribb et al.* 11202! & Issongo, 3 Feb. 1932, *Schlieben* 1718!
DISTR. **T** 6; not known elsewhere
HAB. Grassland with scattered trees, rocky places; 1000–1100 m.

NOTE. A virtually unknown, presumably very local species.

26. **C. sp. E**

Shrub 2 m. tall or tree to 15 m.*; young stems densely shortly pubescent with ± spreading hairs; older stem slender, blackish, ridged and lenticellate, roughened with old petiole-base scars. Leaf-blades elliptic, 1.8–13.5 cm. long, 1.2–6.8 cm. wide, shortly abruptly acuminate at the apex, ± rounded at the base, discolorous, blackish above and covered with adpressed slightly rough white hairs and similar softer hairs beneath; petiole 0.6–3 cm. long, densely shortly pubescent. Cymes aggregated into a terminal panicle 7 cm. long; peduncle 4.5 cm. long; axes densely shortly pubescent. Mature flowers not seen; buds clavate, ± 5 mm. long, very shortly apiculate, densely adpressed pubescent. Fruits oblong-ovoid, 1.4 cm. long, 7 mm. wide, sharply beaked; calycine cup 8 mm. tall, 1–1.2 cm. wide, venose and puberulous.

---

* Field-note to *Vollesen* in *MRC* 4411! (DSM!) states tree to ± 5 m. high.

Tanzania.   Ulanga District: Msolwa Forest, 6 Feb. 1977, *Vollesen* in *MRC* 4411!; Iringa District:
Kidatu, 27 Mar. 1971, *Mhoro* 837!
Distr.   T 6, 7; not known elsewhere
Hab.   Lowland evergreen forest; 250–415 m.

Syn.   *C. sp.* sensu Vollesen in Opera Bot. 59: 75 (1980)

Note.   Vollesen reports 'This can not be matched with any African species and seems to be closest
to the Indian *C. grandis* Roxb. It has the ample inflorescence of *C. millenii* Bak. and *C. platythyrsa*
Bak. but quite different leaves and indumentum'.

27.   C. sp. F

Shrub; stems grey, lenticellate, longitudinally rugulose and with scattered nodular
petiole-bases. Leaf-blades ? oblong, scarcely developed, 1.5+ cm. long, 1.4 cm. wide,
mucronulate at the apex, cuneate at base, glabrous save for a few hairs on midrib; petiole
± 8 mm. long. Inflorescence at apices of shoots in very young bud, 3 cm. long including 1.5
mm. long peduncle; bracts oblong-spathulate, 5 mm. long, ferruginous pubescent; bracts
at base of peduncles sitting on apices of previous growth, similarly pubescent; buds
glabrous, very young but apparently ♂ with rudimentary ovary lacking style.

Tanzania.   Kilosa District: Kondoa–Kimamba, 30 Nov. 1925, *Peter* 45604!
Distr.   T 6; known only from this gathering
Hab.   Forest ('Wald'); 480 m.

Note.   Material inadequate but not matched with any known species.

28.   C. africana *Lam.*, Tab. Encycl. 1: 420 (1792); E.P.A.: 766 (1961); Heine in F.W.T.A.,
ed. 2, 2: 320, fig. 276 (1963); Taton, F.C.B., Borag.: 4, map 1 (1971); Heine in Fl.
Nouv.-Caléd. 7: 106 (1976); Geerling, Guide Terr. Lign. Sah. Soud.-Guin.: 83, t. 17/1
(1982); Troupin, Fl. Rwanda 3: 259, fig. 84/1 (1985); Warfa, Acta Univ. Upsal. 174, I: 1
(1988) & in Taxon 37: 961, fig. 1 (1988); Martins in F.Z. 7(4): 61 (1990). Type: "wanzey",
Ethiopia, Bruce, Travels, 5 (Appendix): 54, t. 17 (1790) (lecto.!)

Shrub or usually a small tree to about 10 m. but occasionally much taller and attaining
18 or even 30 m.; bole usually curved or crooked, 3–8(–12) m. high; bark dark to pale
brown, rough and fibrous, peeling, longitudinally fissured; inner bark white turning
black; slash yellow or white turning green, grey or brownish; crown spreading, rounded
or umbrella-like. Young stems brown-velvety tomentose and with longer hairs but later ±
glabrous. Leaf-blades alternate, ovate, elliptic or almost round, 7.5–17.5(–30) cm. long,
3.5–9(–22.5) cm. wide, rounded to acuminate at the apex, rounded to cordate or rarely
cuneate at the base, ± thick, discolorous, glabrous to scabrid above, shortly tomentellous
and with longer pubescence on the raised reticulate venation beneath to very thickly
velvety ('*C. holstii*') or eventually ± glabrous save for long hairs on the venation; petiole
1.3–13 cm. long. Flowers ♂, sweetly scented, sessile, hairy, having much the appearance of
artificial flowers made of crinkly paper, massed in compact conspicuous panicles of
scorpioid cymes up to 14 × 14 cm., shorter than the leaves. Calyx tubular-campanulate,
(5–)7–9 mm. long, 4–6 mm. wide at the throat, conspicuously 10-ribbed, dark-brown
tomentose, irregularly 4–5-toothed or sometimes ± 2-lipped. Corolla white, funnel-
shaped, spreading at the apex, (1.4–)2–2.5 cm. long and about as wide; tube mostly 1.8–2.2
cm. long and lobes usually 2–3 mm. long, 1.1 cm. wide, retuse, strongly folded, crenellate,
sparsely pubescent outside, the median nerves with a short hairy mucro. Stamens
included; anthers blackening; filaments 8–10 mm. long. Ovary ovoid, 2–3 mm. long,
1.5–1.8 mm. wide, glabrous; style 1.2–2 cm. long, divided 0.7–1.4 cm. from the base;
stigmas subulate, 1–2 mm. long. Fruits yellow, subglobose or ovoid, 1–1.2 cm. long, 6–10
mm. wide, with sweet mucilaginous flesh, glabrous, contained in the slightly accrescent
cupuliform still 10-lobed calyx ± 1.1 cm. wide; endocarp subquadrangular; seeds 2–4. Figs.
5/11, 12, p. 24; 8, p. 32.

Uganda.   W. Nile District: Koboko, Aug. 1938, *Hazel* 666!; Kigezi District: Kinkizi, Kirima, Oct. 1947,
Purseglove 2543!; Mengo District: 6.4 km. on Kampala–Jinja road, July 1932, *Eggeling* 464!
Kenya.   Northern Frontier Province: Marsabit, 13 June 1960, *Oteke* 174!; Baringo District: Kamasia,
Tarambas Forest, Nov. 1930, *Dale* in *F.D.* 2456!; Nyeri District: 7 km. on Nyeri–Kiganjo road, 20 May
1974, *Faden et al.* 74/580!
Tanzania.   Moshi District: Lyamungu, 23 Aug. 1932, *Greenway* 3104!; Lushoto District: W. Usambara
Mts., Bumbuli–Mazumbai road, 8 May 1953, *Drummond & Hemsley* 2433!; Buha District: Gombe
Stream Reserve, Kasakela valley, 13 Apr. 1964, *Pirozynski* 681!

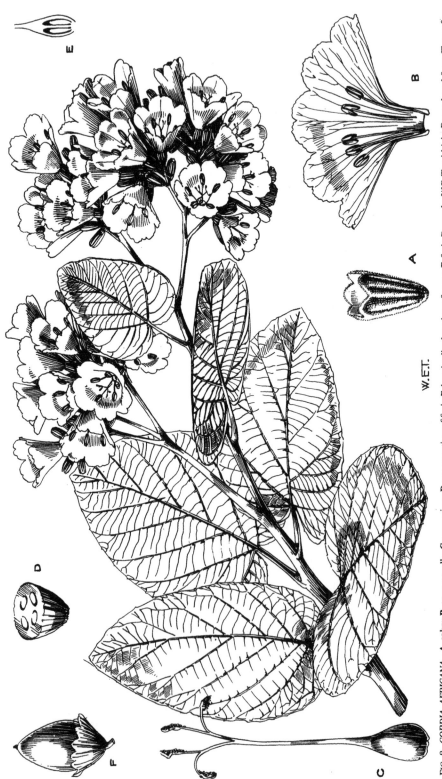

FIG. 8. *CORDIA AFRICANA*—A, calyx; B, open corolla; C, gynoecium; D, cross-section of fruit; E, longitudinal section of ovary; F, fruit. Drawn by W.E. Trevithick. Reproduced from Flora of West Tropical Africa.

W.E.T.

DISTR.   U 1–4; **K** 1–5; **T** 1–7; Guinée to Ethiopia and Sudan thence to Malawi, Mozambique, Zambia, Zimbabwe, South Africa (N. Transvaal) and Angola, also in tropical Arabia; cultivated in India, New Caledonia and elsewhere in tropical botanic gardens, etc.

HAB.   Primary forest edges, secondary forest, gallery forest; also in grassland with scattered trees, *Combretum, Albizia*, etc.; 450–2100 m.

SYN.   *Cordia sebestena* L. var. ß Poir. in Encycl. Méth. Bot. 7: 45 (1806). Type: as for *C. africana*
     C. abyssinica R.Br. in Salt, Voy. Abyss., App.: lxiv (1814); A.Rich., Tent. Fl. Abyss. 2: 80 (1851); Bak. & Wright in F.T.A. 4(2): 8 (1905); Z.A.E. 2: 279 (1911); F.P.N.A. 2: 126 (1947); T.T.C.L.: 75 (1949); I.T.U., ed. 2: 46, pl. 3, fig. 11 (1951); Wimbush, Cat. Kenya Timbers: 38 (1957); Cavaco in Publ. Cultur. Comp. Diamantes Angola 42: 133 (1959); K.T.S.: 69, fig. 15, t. 6 (1961); F.F.N.R.: 364 (1962); White in Chapman & White, Evergreen Forests of Malawi: 182 (1970); White, Veg. of Africa: 333 (1983). Type as for *C. africana*
     Varronia abyssinica (R.Br.) DC. & A.DC., Prodr. 9: 469 (1845)
     Cordia abyssinica R.Br. var. *acutifolia* A.Rich., Tent. Fl. Abyss. 2: 81 (1850). Type: Ethiopia, none stated
     C. holstii Gürke in P.O.A.C.: 335, t. 41 (1895); Z.A.E. 2: 279 (1911); Staner in Rev. Zool. Bot. Afr. 23(2): 224 (1933); Jex Blake, Gard. E. Afr., ed. 4: 221 (1957). Types: Zaire, Lendu Plateau, *Stuhlmann* 2735 (B, syn.†) & Tanzania, W. Usambara Mts., Kwa Mshusa, Kigara, *Holst* 2347 (B, syn.†, K, isosyn.!) & *Holst* 9068 (B, syn.†, K, isosyn.!) & Moshi District, Marangu, *Volkens* 249 (B, syn.†, K, isosyn.!)
     C. unyorensis Stapf in J.L.S. 37: 527 (1906). Type: Uganda "Unyoro, Bugoma and Budongo Forest"*, *Dawe* 798, pro parte, excl. fruit (K, lecto.!)
     Calyptrocordia abyssinica (R.Br.) Friesen in Bull. Soc. Bot. Genève, sér. 2, 24: 180 (1933)

NOTE.   One of the best known timber trees usually called by its Kikuyu name 'muringa', extensively used for drums and bee-hives, also canoes, gongs, mortars, etc. Wood brown, easy to work, remaining stable under changes of humidity, warping little, taking a good polish and thus useful for cabinet making; also used as coffee shade.

     Since there is likely to be an ongoing argument about the correct name for this species a brief summary of the facts is given here. Lamarck cited three syntypes, one with a query so it can be neglected (Alpini reference); the reference to Bruce's excellent plate of 'wanzey' has been selected by Cufodontis as the lectotype so according to the rules there is no doubt that Lamarck's name is correct unless the lectotypification is overthrown. The other reference "ex Africa Lippi herb apud D. Juss." refers to a specimen in the Jussieu Herbarium No. 6475; this still extant herbarium specimen agrees well with Lamarck's short description and is *Cordia myxa* L.; in any case Lippi did not collect anywhere within the range of the species described above. F. White argues that *C. africana* Lam. is a synonym of *C. myxa* and that the citation of Bruce's plate is an error. Heine considers that Lamarck was too good a botanist to deal with *C. myxa* twice under different names since it is treated under its own name further on; also Poiret deals with *C. myxa* separately. Both *C. africana* and *C. abyssinica* are likely to be used for the species in question for many years.

2. EHRETIA

P.Br., Civ. Nat. Hist. Jam.: 168 (1756)

Trees or shrubs with alternate, petiolate, simple, entire or toothed leaves. Flowers mostly small, ⚥ or polygamous, sessile or pedicellate in much branched axillary or terminal cymose inflorescences, the individual cymules mostly scorpioid. Calyx mostly small, campanulate, deeply divided into 4–5 imbricate lobes, ciliate but glabrous inside; tube short. Corolla white, yellowish or blue, campanulate or subrotate, usually small, the tube very short or longer than the lobes; lobes (4–)5(–7), oblong, imbricate, spreading or reflexed. Stamens inserted at or near the throat, usually exserted; anthers oblong. Ovary subglobose with either 2, 2-ovuled locules or 4, 1-ovuled locules; style terminal, scarcely to deeply divided into 2 branches terminated by capitate stigmas. Fruits subglobose, fleshy, breaking up at maturity into 2, 2-locular mericarps or 4, 1-locular mericarps. Seeds 4 or less by abortion.

     About 75 species in the tropics of both Old and New Worlds, particularly in Africa and Asia; a few in tropical America and West Indies. *Ehretia tetrandra* Gürke described from Tanzania, Usagara, Ruaha R., is actually *Premna senensis* Klotzsch.

1. Inflorescence-axes and leaves glabrous or almost so; corolla-tube (3–)4–8 mm. long   2
    Inflorescence-axes and leaves pubescent to pilose or velvety   3

* Kew sheet labelled Budongo Forest

2. Leaf-blades oblong to elliptic, 1–18 × 0.5–10.5 cm.; flowers
      mostly more or less precocious; corolla-tube (3–)4–5.5
      mm. long . . . .   5. *E. bakeri*
    Leaf-blades obovate, 0.6–6 × 0.35–4 cm.; flowers and
      mature leaves usually occurring together; corolla-tube
      6.5–8(–10) mm. long . . . .   3. *E. janjalle*
3. Inflorescence-axes densely glandular-pubescent; corolla
      pink to purple 4
    Inflorescence-axes pubescent to pilose but not glandular;
      corolla white, yellowish or pinkish white 5
4. Leaves pubescent to velvety; calyx-lobes ovate, elliptic
      or narrowly triangular, 2 mm. long (widespread)   2. *E. obtusifolia*
    Leaves glabrous; calyx-lobes subulate from a triangular
      base, acuminate, 2.5–3 mm. long (T8) . . . . .   4. *E. glandulosissima*
5. Leaves ± rough above with tubercle-based short hairs and ±
      velvety beneath, the venation not impressed above;
      mostly a woodland or bushland species occurring up to
      600 m. . . . .   1. *E. amoena*
    Leaves glabrous save for midrib above and nervation and
      domatia beneath, often somewhat bullate with venation
      ± impressed above; mostly a forest species occurring
      from 1000–2400 m. . . . .   6. *E. cymosa*

1. **E. amoena** *Klotzsch* in Peters, Reise Mossamb. Bot.: 248, t. 41 (1861); Gürke in P.O.A.
C: 335 (1895); C.H. Wright in Fl. Cap. 4(2): 5 (1904); Bak. & Wright in F.T.A. 4(2): 24 (1905);
Brenan in Mem. N.Y. Bot. Gard. 9: 4 (1954); K.T.S.: 72 (1961); Palgrave, Trees S.Afr.: 802
(1977); Vollesen in Opera Bot. 59: 75 (1980); Martins in F.Z. 7(4): 76 (1990). Type:
Mozambique, surroundings of Sena, *Peters* (B, holo., K, iso.!)

Shrub or small tree 1.8–7.5(?–12) m. tall with ashy white glabrous branches; bark rather
rough, grey; young branches at first shortly pubescent. Leaf-blades obovate-elliptic to
oblong or ± round, 2–13(–20) cm. long, 1–11.5(–14.5) cm. wide, obtuse, emarginate or very
obscurely acuminate at the apex, cuneate to rounded or slightly subcordate at the base,
somewhat rough above with tubercle-based short hairs and almost velvety beneath with
rather similar hairs, the venation ± raised beneath, entire to vaguely undulate; petiole
1.2–1.5 cm. long. Flowers slightly scented, in extensive inflorescences up to 10–15 cm.
long, 11–16 cm. wide; peduncles 1.5–3 cm. long; pedicels 0.5–1.5 mm. long; inflorescence-
axes densely pubescent with long and short hairs. Calyx 2 mm. long including narrowly
triangular lobes 1.5 mm. long, 0.8 mm. wide, densely glandular-pubescent. Corolla white;
tube 1.5–2 mm. long; lobes triangular-oblong, 3 mm. long, 1.3 mm. wide, reflexed.
Stamens exserted; anthers purplish; filaments 2 mm. long; style exserted, 2–3 mm. long
with branches 0.75–1.2 mm. long. Fruits yellow, becoming scarlet when ripe, globose, 3–5
mm. diameter in dry state, wrinkled; pyrenes ¼-spherical, 4 mm. long, 3.5 mm. wide, one
inner face deeply excavated, outer rounded one obscurely ridged.

KENYA. Kilifi/Kwale Districts: Mazeras, *R.M. Graham* 414 in *F.D.* 1718! & 14 Mar. 1902, *Kassner* 282!;
      Mombasa, Sept. 1873, *Kirk*!
TANZANIA. Lushoto District: Korogwe–Handeni road, 3 Mar. 1954, *Faulkner* 1367!; Morogoro, 24
      Dec. 1934, *E.M. Bruce* 369!; Lindi District: 8 km. W. of Mtama near R. Nyango, 14 Dec. 1955,
      Milne-Redhead & Taylor 7488!
DISTR. **K** 7; **T** 3, 6, 8; Mozambique, Malawi, Zambia, Zimbabwe, Swaziland and South Africa (Natal
      and Transvaal)*
HAB. Woodland, bushland and thicket-clumps including *Brachystegia* and *Albizia* woodland, *Acacia*
      thicket, etc., often on termite mounds; 0–600 m.
SYN. *E. stuhlmannii* Gürke in P.O.A. C: 336 (1895) & in E.J. 28: 309 (1900); Bak. & Wright in F.T.A. 4(2):
      27 (1905); T.T.C.L.: 77 (1949). Types: Tanzania, Uzaramo District, Dar es Salaam, *Stuhlmann*
      57 (B, syn.†) & Nyika, Mtindi Market to Kwa Mkembe, *Volkens* 35 (B, syn.†)
    E. goetzei Gürke in E.J. 28: 311, 461 (1900); Bak. & Wright in F.T.A. 4(2): 26 (1905); T.T.C.L.: 77
      (1949). Type: Tanzania, Morogoro District, Ukutu [Khutu], Mgeta R., near Kisaki, *Goetze* 128
      (B, holo.†, K, iso.!)
    [*E. corymbosa* sensu Fosberg in K.B. 29: 260 (1974), pro parte, *non* A.DC.]

---

* The identity of the two specimens I have seen from the Sudan (F.P.S. 3: 81 (1956)) is uncertain.

NOTE. Brenan (loc. cit.) suggests Baker and Wright were wrong to synonymise *E. mossambicensis* with the above plant and that more likely the name belongs to the next species. Since in neither case would it interfere with the names and no type is now available the matter is academic. It seems remarkable that the only material available from Kenya is little more than Baker & Wright had for F.T.A. *E. amoena* certainly has nothing to do with *E. corymbosa* as claimed by Fosberg; the claim is not repeated in Fl. Aldabra (1980). Martins (F.Z. 7(4): 79 (1990)) notes many *E. amoena/E. obtusifolia* intermediates in the Flora Zambesiaca area.

2. **E. obtusifolia** *A.DC.* in DC., Prodr. 9: 507 (1845); A.Rich., Tent. Fl. Abyss. 2: 83 (1850); C.B. Cl. in Fl. Brit. Ind. 4: 142 (1883); Hiern, Cat. Afr. Pl. Welw. 1: 716 (1898); E.P.A.: 771 (1961); Martins in F.Z. 7(4): 78 (1990). Type: Ethiopia, Tigre, Medschara, near Gapdia, *Schimper* 652 (G, holo., K, iso.!)

Shrub or small tree, usually virgately branched or small often straggling tree, often with several branches from the base, 1.8–6 m. tall, deciduous; bark grey; young shoots glandular-pubescent and with white non-glandular hairs and some adpressed pubescence, sometimes ferruginous; older stems pale brown, ± striate. Leaf-blades elliptic, oblong-elliptic or obovate, 1.2–9(–11) cm. long, 0.7–6(–8.5) cm. wide, obtuse to shortly acuminate at apex but sometimes truncate, rounded or even ± emarginate, cuneate at the base, pubescent to velvety on both surfaces but not usually rough, the hairs on the upper surface, however, slightly bulbous-based but occasionally quite roughish ('rougher with age' *fide B.D. Burtt* 559); petiole up to 2 cm. long. Flowers sweet-scented, in glandular pubescent terminal cymes 2.5 cm. wide but sometimes on very short lateral shoots and appearing axillary; where cymes are close on leafless branches the flowers are in dense masses up to 30 cm. long; peduncles 0.5–3 cm. long; pedicels 1–4.5 mm. long. Calyx-lobes ovate, elliptic or narrowly triangular, 2 mm. long, 0.8 mm. wide, ciliate and puberulous, joined only at extreme base. Corolla pink, blue, pale lilac or mauve, drying brown and white; tube cylindrical, 4–4.5 mm. long; lobes narrowly triangular to oblong, 3–4.5 mm. long, 1–1.2 mm. wide, acute or obtuse. Filaments exserted 2–4 mm. Style 4–6 mm. long with branches 1.5–2 mm. long or in T4 material 3 mm. long with 3 mm. long branches. Fruits orange or red, globose, 5–6 mm. diameter, glabrous or densely shortly glandular-pubescent (see note); pyrenes subovoid, 4.5 mm. long, 3.5 mm. wide, excentrically excavated on inner two faces so as to appear like a human ear, the excavation nearest to the apical pointed end of the pyrene; outer face convex, very irregularly reticulate.

UGANDA. Karamoja District: Amudat, Nov. 1941, *Dale* U183! & Lokichar R., July 1958 *J. Wilson* 464!
KENYA. Baringo District: 22.9 km. on Loruk–Tengulbei road, 3 Dec. 1989, *Luke B.F.F.P.* 585! & 38.2 km. on same road, 3 Dec. 1989, *Luke B.F.F.P.* 623!
TANZANIA. Mbulu District: Lake Manyara National Park, 29 Oct. 1962, *Dingle* 352!; Singida District: Singida to Kiraminu, 16 Dec. 1933, *Michelmore* 831!; Mbeya District: Trekimboga track, 20 Oct. 1970, *Greenway & Kanure* 14593!
DISTR. U 1; **K**3, ? 7 (Lake Chala) (see note); **T** 1, 2, 4, 5, 7; Ethiopia, Somalia, Mozambique, Malawi, Zambia, Zimbabwe, Botswana, Angola, Namibia and South Africa (Transvaal); also Arabia, Socotra, Pakistan (Sind, Baluchistan, Punjab), NW. India
HAB. *Brachystegia-Uapaca* woodland, riparian woodland, wooded grassland and deciduous thicket and thornbush, often in rocky places; 700–1500 m.
SYN. *E. obovata* R.Br. in Salt, Voy. Abyss. App.: lxiv (1814), *nomen*
    ?*E. mossambicensis* Klotzsch in Reise Mossamb., Bot.: 249, t. 42 (1861); Gürke in P.O.A. C: 335 (1895). Type: Mozambique, Tete, Rios de Sena, *Peters* (B, holo.†)
    E. fischeri Gürke in P.O.A. C: 336 (1895) & in E.J. 28: 313 (1900); Bak. & Wright in F.T.A. 4(2): 27 (1905); T.T.C.L.: 78 (1949). Types: Tanzania, Mwanza District, Kagehi [Kageyi] (= Kayenzi, 2°23'S, 33°06'E), *Fischer* 323 (B, syn.†) & Biharamulo/Mwanza Districts, Usinga (?Uzinza, 3°00'S, 32°00'E), near the French Mission at Usambiro, *Stuhlmann* 850 (B, syn.†)
    E. caerulea Gürke in E.J. 28: 312, 461 (1900); Bak. & Wright in F.T.A. 4(2): 24 (1905), pro parte; T.T.C.L.: 77 (1949). Type: Tanzania, Iringa District, Lukose R., *Goetze* 484 (B, holo.†, K, iso.!)
    E. sp. sensu I.T.U., ed. 2: 50 (1952)
NOTE. *Gilbert* 4896 (**K**7/**T**2) Lake Chala, 2 Nov. 1969, appears to be this species. Material from much of Tanzania and some parts of the Flora Zambesiaca area has the fruits densely shortly glandular-pubescent in contrast to the majority of specimens in which they are glabrous and the style and branches are also ± equal in length (e.g. *Richards & Arasululu* 26298, Iringa District, Ruaha National Park, Kinyantupa [Kinyantubi]); possibly this represents a distinct subspecies but information is inadequate. I have seen no authentic material of *E. fischeri* but the description agrees well with T1 material of the present species. Martins (F.Z. 7(4): 79 (1990)) reports many intermediates in the Flora Zambesiaca area between *E. obtusifolia* and the closely related S. African *E. rigida* (Thunb.) Druce.

3. **E. janjalle** *Verdc.*, sp. nov., *E. obtusifoliae* DC. et *E. braunii* Vatke valde affinis sed foliis glabris vel obscure ciliatis, inflorescentiae axibus glabris, tubo corollae 6.5–8(–10) mm. longo differt. Type: Kenya, Northern Frontier Province, Dandu, *Gillett* 12583 (K, holo.!, BM, EA, K, iso.!)

Shrub to 3 m., branching from the base; young shoots glabrous, older pale grey to grey-brown with numerous pale lenticels; very young unformed leaves with a few scurfy hairs. Leaf-blades obovate to oblong-elliptic, 0.6–6 cm. long, 0.35–4 cm. wide, emarginate to broadly rounded at the apex, cuneate at the base, glabrous or very obscurely ciliate; petiole 0–7 mm. long. Flowers in terminal and axillary divaricate cymes up to 5 × 6 cm.; peduncle and secondary branches up to ± 1 cm. long; pedicels 0.5–3 mm. long; all axes glabrous. Calyx-lobes ovate to ovate-triangular, 2–2.5 mm. long, 1.8–2.5 mm. wide, rounded at apex, ciliolate, pubescent inside. Corolla pale purple, drying brown with white base to the tube; tube cylindrical, 6.5–8(–10) mm. long; lobes 3.2–4 mm. long, 1.6–2.5 mm. wide, sparsely ciliate. Filaments exserted, 2 mm. long; anthers 1.3 mm. long. Style 5–6.5 mm. long including the 1.5–1.8 mm. long branches; stigmas ± capitate. Fruits orange, depressed-globose, 4.5–6 mm. diameter; pyrenes segment-shaped, 4 mm. long, ± 3 mm. wide, one inner face with ear-shaped excavation, the other smooth without opening, outer face rugulose, reticulately pitted.

KENYA.   Northern Frontier Province: Dandu, 18 Mar. 1952 (fl.) & 11 Apr. 1952 (fr.), *Gillett* 12583!
DISTR.   **K** 1; Ethiopia, S. Somalia
HAB.   *Acacia-Commiphora* bushland, rocky places near water-holes; 800 m.

NOTE.   A precocious plant from S. Somalia (*Becket & White* 1781) with at least the leaf-margins bristly asperulous and corolla up to 1 cm. long seems conspecific. *Hucks* 1028 (Kenya, Teita District, 8 km. on Kima–Mudanda road, 31 May 1967) described as a 4.5 m. bush with red berries, has only fruiting calyces and some pyrenes remaining but is probably this species as indicated by Greenway (adnot.) who equates it with *Gillett* 12583; it differs in the leaf-venation being impressed above and prominent beneath.

4. **E. glandulosissima** *Verdc.*, sp. nov., affinis *E. obtusifoliae* DC. sed foliis glabris, lobis calycis basi triangularibus apice subulato-acuminatis ± 3 mm. longis differt. Type: Tanzania, Lindi District, Rondo Plateau, Mchinjiri, *Bryce* 16 (K, holo.!, EA, TFD, iso.!)

Small tree 4.5–6 m. tall; young shoots glabrous but leaf-buds sticky; bark buff, wrinkled and striate. Leaf-blades elliptic-oblong, 6.5–11 cm. long, 3.5–5.5 cm. wide, shortly acuminate at the apex, cuneate at the base, glabrous save for a few hairs in the axils on domatia beneath; petiole 1.2–1.5 cm. long. Flowers in many-flowered much-branched cymes 6.5 cm. long, 7.5 cm. wide, with short peduncles and secondary peduncles ± 1 cm. long; pedicels 1–6 mm. long; axes densely covered with short ferruginous glandular hairs. Calyx 4 mm. long with similar indumentum to the axes; lobes triangular at base, 2.5–3 mm. long, 0.7 mm. wide, subulate-acuminate. Corolla purple; tube 4–5 mm. long; lobes oblong, 4 mm. long, 2.2 mm. wide. Stamens exserted from tube for ± 3 mm. but not exceeding the lobes. Style 8 mm. long including 2.5–3.5 mm. long arms; stigmas not capitate. Fruit not known.

TANZANIA.   Lindi District: Rondo Plateau, Mchinjiri, 1 Oct. 1951, *Bryce* 16!
DISTR.   **T** 8; not known elsewhere
HAB.   Moist evergreen forest; 810 m.

NOTE.   The material is very inadequate, only 1 inflorescence and 1 broken leaf being preserved at Kew; 2 rather better leaves are present on the TFD specimen. The tie-on label on the Kew specimen bears a native name mmangu (or similar) which may help to locate further material.

5. **E. bakeri** *Britten* in J.B. 33: 88, adnot. (1895); Bak. & Wright in F.T.A. 4(2): 22 (1905); T.T.C.L.: 76 (1949); K.T.S.: 72 (1961). Type: Tanzania, Uzaramo District, Dar es Salaam, *Kirk* 124 (K, holo.!)

Shrub or small deciduous tree 1.8–6 m. tall; bark grey; young branches whitish grey, lenticellate, glabrous or youngest parts ± ferruginous tomentose; bark and flowers said to contain white latex (*fide* G.H. Donald in F.D. 2501). Leaves flushed bronze when very young; blades oblong to elliptic, 1–18 cm. long, 0.5–10.5 cm. wide, rounded to shortly acuminate at the apex, broadly cuneate to subcordate at the base, glabrous above, minutely pubescent beneath, or young leaves finely tomentose on both sides; occasionally persistently pilose on both sides; venation finely raised beneath; petiole up to 2 cm. long. Flowers slightly scented, in ± dense many-flowered inflorescences 2–4 cm.

long, 3–6.5 cm. wide with ?glutinous axes, mostly on leafless branches; peduncles and secondary peduncles ± 1 cm. long; pedicels 1–1.5 mm. long. Calyx 1.8–2(–3) mm. long, deeply divided into ovate-lanceolate to ± triangular-subulate minutely ciliate lobes. Corolla white or pale pinkish mauve; tube cylindrical, 4–5.5 mm. long; lobes narrowly triangular or ovate, 2–3.5 mm. long, 1–1.5 mm. wide, reflexed, ± finely ciliate. Anthers yellow to brown, exserted; filaments 3–4.5 mm. long. Style 3.5–5 mm. long with branches 1–2 mm. long. Fruit globose, 5 mm. diameter in dry state; pyrenes brown, segment-shaped, 4 mm. long, 3 mm. wide, almost entirely hollow, the seed restricted to a narrow area; outer surface with ± 2 ridges and obscure reticulation, one inner face open, the third impressed with a serrated ridge where it meets the outer surface.

KENYA. Teita District: Mbololo Hill, *Gardner* in *F.D.* 2986!; Kwale District: Taru, 8 Sept. 1953, *Drummond & Hemsley* 4201!; Lamu District: 48 km. N. of Lamu, Lungi Forest, Oct. 1937, *Dale* in *F.D.* 3825!
TANZANIA. Moshi District: Pare Mts., Mugiligili [Mgigile], Oct. 1927, *Haarer* 820!; Pangani District: 29 km. SW. of Pangani, Msubugwe Forest, 17 Mar. 1950, *Verdcourt* 119!; Morogoro, Nov. 1951, *Eggeling* 6350!; Pemba I., Kangagaani, Sept. 1928, *Vaughan* 386!
DISTR. **K** 7; **T** 2, 3, 6, 8; **P**; not known elsewhere
HAB. *Pterocarpus, Pleurostylia, Brachystegia, Sclerocarya* thicket forest, thicket on shallow soil over rock or on coral rag, bushland, inner edges of mangrove swamps and gneiss slopes; 0–1050 m.

SYN. *E. macrophylla* Bak. in K.B. 1894: 29 (1894), *non* Wall. Type: as above

6. **E. cymosa** *Thonn.* in Schumach. & Thonn., Beskr. Guin. Pl.: 129 (1827); Bak. & Wright in F.T.A. 4(2): 25 (1905); Brenan in Mem. N.Y. Bot. Gard. 9(1): 5 (1954); Aubrév., Fl. For. Côte d'Ivoire, ed. 2, 3: 217, t. 332/69 (1959); Heine in F.W.T.A., ed. 2, 2: 318 (1963); Taton, F.C.B., Borag.: 22 (1971); Riedl in Linzer Biol. Beitr. 17: 303–315 (1985).* Type: Ghana [Guinea], *Thonning* 89 (C, holo. & iso., FT, P-JU, iso.)

Shrub or small tree 2–9(–20) m. tall, with spreading crown and often weak drooping branches. Young stems pubescent or glabrescent, soon glabrous. Leaf-blades elliptic or elliptic- to ovate-oblong, 7.5–20 cm. long, 3.5–12 cm. wide, acuminate at the apex, cuneate, rounded or ± subcordate at the base, glabrous above save for the midrib and beneath save for pubescent nervation and hairy domatia, ± subcoriaceous; venation ± impressed above, raised beneath; petiole 1.2–3.5 cm. long, glabrous, puberulous or hairy. Flowers ⚥, fragrant, in often copious panicles of corymbose cymes up to 15 cm. long and wide, the axes glabrous to pubescent or long densely spreading hairy, sessile or subsessile or terminal flowers shortly pedicellate; pedicels articulate at the base, 1–2(–3) mm. long. Calyx campanulate, 1.5–2 mm. long, with 4–5 lanceolate lobes 0.7–1.2 mm. long, acute, glabrous or puberulous, ciliate. Corolla white, yellowish or pinkish white, campanulate, 4–8 mm. long; tube 1.5–3 mm. long; lobes 4–5, oblong, 1.5–4 mm. long, 1.5–1.8 mm. wide, obtuse or subacute, reflexed, glabrous or sparsely to densely ciliate at the apex. Stamens exserted; filaments 2.4–3.5 mm. long; anthers brownish or blackish. Ovary ovoid, 1 mm. long, glabrous; heterostyly frequent; style white or pale purple, exserted 2–4 mm. in long-styled flowers, 0.7–0.8 mm. long and included in short-styled flowers; stigmatic branches 0.5–1 mm. long. Fruits orange, red or black, ovoid to subglobose, 2–6 mm. in diameter, apiculate, breaking up at maturity into 4 mericarps; pyrenes subovoid, 3.3 mm. long, 2.8 mm. wide; one inner face with an excavation towards the apex, the other not excavated but with an indentation bordering the straight margin; outer face convex with 4–5 ribs strongest towards the base and smooth towards the right-hand margin.

KEY TO INFRASPECIFIC VARIANTS

1. Calyx 1–1.5 mm. long; corolla-lobes 1.5–2.5 mm. long;
   inflorescence-axes puberulous or shortly pubescent   a. var. **cymosa**
   Calyx 2–2.5 mm. long; corolla-lobes 2.5–3.5 mm. long 2
2. Inflorescence-axes puberulous or shortly pubescent   b. var. **divaricata**
   Inflorescence-axes long-pubescent or long-pilose   c. var. **silvatica**

a. var. **cymosa**

Inflorescence-branches very shortly pubescent or puberulous. Flowers usually all or nearly all distinctly pedicellate. Calyx 1–1.5 mm. long. Corolla-lobes 1.5–2.5 mm. long.

* *E. cymosa* Roem. & Schultes (1819) was not validly published.

UGANDA. Bunyoro District: Budongo Forest, Feb. 1935, *Eggeling* 1614! & 1628!; Kigezi District: Maramagambo Forest, Sept. 1936, *Eggeling* 3320!; without locality, *Dawe* 1005!
DISTR.   U 2, 4; West Africa from Sierra Leone to Gabon, Zaire
HAB.   Evergreen forest; ? 1000–1200 m.

SYN.   *E. thonningiana* Exell, Suppl. Cat. Vasc. Pl. S. Tomé: 34 (1956). Type: as for *E. cymosa*
      [*E. corymbosa* sensu Fosberg in K.B. 29: 260 (1974), pro parte, ?*non* A.DC.]

NOTE.   I am not at all sure that the Aldabra plant dealt with by Fosberg has anything at all to do with
      E. cymosa, which name is not in any case displaced by Roemer & Schultes' earlier but invalid one.

b. var. **divaricata** (*Bak.*) *Brenan* in Mem. N.Y. Bot. Gard. 9(1): 5 (1954); K.T.S.: 72, adnot. (1961); Palgrave, Trees S. Afr.: 803 (1977); Martins in F.Z. 7(4): 75 (1990). Type: Malawi, Chiradzulu Mt., *Kirk*, (K, holo.!)

Inflorescence-branches very shortly pubescent or puberulous. Flowers, save the lowest, sessile or subsessile. Calyx 2–2.5 mm. long. Corolla ± 4.5–5.5 mm. long, the tube 2 mm. and lobes 2.5–3.5 mm. long.

UGANDA.   Toro District: Hima [Hema], 6 Sept. 1905, *Dawe* 506!; Kigezi District: Kanungu, Apr. 1948, *Purseglove* 2632! & Lake Chahafi, 12 Jan. 1937, *Rogers & Gardner* 537!
KENYA.   Northern Frontier Province: Mt. Kulal, 7 June 1960, *Oteke* 53!; "Kikuyu and on way to Eldama Ravine", *Whyte*!; Teita District: Chawia Forest, 22 May 1985, *Beentje et al.* 855!
TANZANIA.   Lushoto District: W. Usambara Mts., Sunga–Manolo road, 26 May 1953, *Drummond & Hemsley* 2782!; Ufipa District: Mbizi Forest, Nov. 1958, *Napper* 1091!; Rungwe District: Kyimbila, perhaps Tandala, Nov. 1913, *Stolz* 2236!; Zanzibar I., Ufufuma, 4 Sept. 1932, *Vaughan* 1958! (see note)
DISTR.   U 2; K 1, 3, ?4, 7; T 1, 3, 4, 6, 7; Z; Zaire, Ethiopia (*fide* Martins), Mozambique, Malawi and Zimbabwe
HAB.   Dry upland *Juniperus* forest and rather wetter evergreen forests, also in derived grassland, bushland and secondary growth on hillsides; (0–)1500–2100(–2400) m.

SYN.   *E. divaricata* Bak. in K.B. 1894: 28 (1894); Bak. & Wright in F.T.A. 4(2): 26 (1905); T.T.C.L.: 77
      (1949)

NOTE.   Intermediates between var. *divaricata* and var. *silvatica* have been found in Ankole and Mubende and on Mt. Kulal. The occurrence at a low altitude on Zanzibar seems extraordinary. The individual cymules of the inflorescence are denser and subglobose but the identity seems clear. The species does of course occur at low altitudes in West Africa. Despite a firm No! annotation by Brenan, *Swynnerton* 2044 (Tanzania, near Lushoto, Magamba Forest, 21 Feb. 1921) belongs here.

c. var. **silvatica** (*Gürke*) *Brenan* in Mem. N.Y. Bot. Gard. 9(1): 5 (1954); K.T.S.: 72 (1961); E.P.A.: 770 (1962); Hepper, Jaeger et al., Annot. Check-list Pl. Mt. Kulal: 86 (1981); Troupin, Fl. Rwanda 3: 262, fig. 84/3 (1985). Types: Tanzania, W. Usambara Mts., Kwa Mshuza, *Holst* 9067 (B, syn.†, K, isosyn.!) & Kilimanjaro, Marangu, *Volkens* 1470 (B, syn.†)

Inflorescence-branches ± densely covered with ± long spreading hairs, rarely ± glabrous. Flowers mostly sessile or subsessile. Calyx 2–3 mm. long. Corolla-lobes 2–3 mm. long. Leaves often quite densely pubescent beneath. Fig. 9.

UGANDA.   Kigezi District: Kachwekano Farm, July 1949, *Purseglove* 2982!; Teso District: Kumi, Feb. 1916, *Snowden* 268!; Masaka District: Kabula, Oct. 1932, *Eggeling* 694!
KENYA.   Mt. Elgon, 11 Apr. 1931, *E.J. & C. Lugard* 554a!; Kiambu District: Muguga, 13 July 1952, *Verdcourt* 682!; N. Kavirondo District: Kakamega Forest, 15 Oct. 1953, *Drummond & Hemsley* 4785!
TANZANIA.   Masai District: Ngorongoro Crater, S. slope above Kampi ya Nyoka, 23 Sept. 1932, *B.D. Burtt* 4288!; Mbulu District: Native Authority Farm, 14 Apr. 1954, *Matalu* 3073!; Moshi District: Kilimanjaro, Machame [Mashami], Mronga, Feb. 1928, *Haarer* 1034!
DISTR.   U 2–4; K 1–6; T 2, 3, 7; Ethiopia
HAB.   Rain-forest, riverine forest, also derived bushland and grassland; 960–2250 m.

SYN.   *E. silvatica* Gürke in E.J. 19, Beibl. 47: 46 (1894) & in P.O.A. C: 336 (1895); Bak. & Wright in F.T.A.
      4(2): 23 (1905); T.T.C.L.: 77 (1949); I.T.U., ed. 2: 50 (1952)
      E. inamoena Standley in Smithsonian Misc. Coll. 68(5): 12 (1917). Type: Kenya, Nyeri to
      Wambugu's, *Mearns* 1981 in US 631940 (US, holo., BM, iso.!)
      E. abyssinica Fresen. var. *silvatica* (Gürke) Riedl in Linzer Biol. Beitr. 17: 313 (1985)

NOTE.   Riedl (Linzer Biol. Beitr. 17: 303–315 (1985)) has discussed Brenan's treatment of *E. cymosa* and disagreed with the division into varieties. I. Friis (pers. comm.) considers two subspecies based on corolla size might be the correct solution.

FIG. 9. *EHRETIA CYMOSA* var. *SILVATICA*—**1**, flowering branch, × ²/₅; **2**, flower, × 6; **3**, calyx, × 10; **4**, corolla, opened up, × 6; **5**, stamen, × 10; **6**, ovary and stigma, × 12; **7**, ovary, transverse section, × 24; **8**, fruit, × 4; **9**, fruit, transverse section, × 5; **10**, pyrene, × 6. 1–7, from *Lugard* 554a; 8, from *Newbould* s.n.; 9, 10, from *Snowden* 651. Drawn by Mrs Maureen Church.

IMPERFECTLY KNOWN SPECIES

7. **E. uhehensis** *Gürke* in E.J. 28: 462 (1900); Bak. & Wright in F.T.A. 4(2): 27 (1905); T.T.C.L.: 78 (1949). Type: Tanzania, Iringa District, Lofia [Lofio] R., *Goetze* 444 (B, holo.†)

Shrub with long overhanging stems bearing dense short downy shoots; bark pale grey. Leaf-blades oblong, 4–7 cm. long, 2.5–4.5 cm. wide, acute at the apex, cuneate at the base, the margin irregularly toothed towards the apex, finely pubescent; petiole 1–2 cm. long. Cymes terminal, very lax, finely pubescent, the axes finely downy pubescent as in the young shoots; bracts lanceolate-subulate, 2–5 mm. long. Calyx tubular, 3 mm. long, 5-lobed to the middle, finely downy pubescent outside; teeth ovate, short, scarcely 1 mm. long, obtuse. Corolla white, tubular, 5 mm. long, divided to half-way into 5 lanceolate obtuse lobes. Anthers exserted. Style 5–6 mm. long, exserted. Fruit not known.

TANZANIA.   Iringa District: Lofia [Lofio] R., Jan. 1899, *Goetze* 444
DISTR.   T 7
HAB.   On grey laterite on mountain slopes, presumably in forest; 600 m.
NOTE.   *Goetze* gave the native name as 'makungu'. Could it perhaps be a *Premna*?

8. **E. rosea** *Gürke* in E.J. 28: 461 (1900); Bak. & Wright in F.T.A. 4(2): 24 (1905); T.T.C.L.: 77 (1949). Types: Tanzania, Morogoro District, E. Uluguru Mts., Tununguo, *Stuhlmann* 8713 & no locality, *Stuhlmann* 8961 (both B, syn.†)

Shrub 1–2 m. tall with grey bark; branches slender, terete, yellowish-grey tomentose. Leaf-blades obovate, 4–6 cm. long, 2–4 cm. wide, obtuse or shortly acuminate, narrowed at the base, leathery, entire, pubescent on both surfaces; petiole 1 cm. long. Cymes lax, 20–30-flowered, 5–7 cm. long; axes glabrous; pedicels 3–5 mm. long. Calyx finely hairy ('downy'); lobes 5, lanceolate, 2 mm. long, 0.5 mm. wide, longer than the tube, acuminate. Corolla rose-red; tube 4 mm. long and "proportionally wide"; lobes 5, lanceolate, 3 mm. long, 1 mm. wide, obtuse. Stamens 5, rather far-exserted. Style 6 mm. long. Fruit 4 mm. (sphalm cm.) diameter.

TANZANIA.   Morogoro District: E. Uluguru Mts., Tununguo, *Stuhlmann* 8713 & without locality, Oct. 1894, *Stuhlmann* 8961
DISTR.   T 6
HAB.   Not known; 170–200 m.
NOTE.   I have been unable to find any authentic material of this and Baker & Wright saw none. Gürke stated that it belonged to sect. '*Beurrerioides*' Benth. & Hook. because of the fruit, but it clearly is not a relative of *Bourreria petiolaris*. He also indicates it is closest to *E. amoena*, which certainly occurs in the general area but has a shorter corolla. It might fit *E. obtusifolia* better but that has not been seen from the area. The glabrous inflorescence axes fit neither.

3. BOURRERIA

P.Br., Civ. Nat. Hist. Jam.: 168, 492 (1756), as "*Beurreria*"; Schulz in Urban, Sym. Antill. 7: 45 (1911), as "*Beurreria*"; Thulin in Nordic Journ. Bot. 7: 413 (1987), *nom. conserv.*

Hilsenbergia Meissn., Pl. Vasc. Gen. Comment.: 198 (1840)

Ehretia P.Br. sect. *Bourreria* A.DC. in DC., Prodr. 9: 504 (1845)

Ehretia P.Br. sect. *Bourrerioides* Benth. in G.P. 2: 841 (1876)

Trees or shrubs, glabrous or tomentose-pubescent. Leaves alternate or subverticellate, obovate-oblong, rarely acuminate, smooth or scabrid above. Flowers small to quite large (up to 5 cm. long), few to many in lax dichotomous corymbose cymes, rarely solitary. Calyx globose or ovoid in bud, closed, splitting into 2–5 valvate teeth or lobes which are sometimes irregular and often cohering, tomentose inside. Corolla-tube short or elongate-cylindrical, often broadened at the throat; lobes 5, ovate or broadly ovate, obovate or cordate, spreading, imbricate. Stamens 5, inserted in the tube, included or exserted; filaments filiform, short; anthers ovate or oblong. Ovary 4-locular; style terminal, bifid, or entire in a few species; stigmas truncate, capitate or clavate; ovules lateral, affixed above or below the middle. Drupes subglobose, small or quite large, up to 2.5 cm. diameter; endocarps ± 4-angled, splitting into 4 bony 3-angled pyrenes with ridged or winged outer faces; central column splitting into 4 erect branches with pyrenes finally suspended from the apices or ± adnate to internal angle of pyrenes or entirely obsolete; seeds laterally affixed; albumen sparse to fairly copious, fleshy.

Some 70 described species from W. Indies, Central and South America with a few in East Africa, Madagascar and the Mascarene Is.; Miers, however, over-elaborated the species and the true number is probably ± 50. It is clear that Schulz (loc. cit. 70 (1911)) thought *Ehretia petiolaris* was a *Bourreria* but he made no new combination. The strange disjunct distribution is mirrored by a number of other genera, e.g. *Hirtella*.

1. Flowers solitary to 2 together or less often in 2(–3)-flowered
cymes; leaves not bullate, small, 0.5–2 × 0.35–1 cm.   3. *B. lyciacea*
Flowers in several–many-flowered inflorescences or, if in
very few-flowered cymes, then leaves ± bullate and
exceedingly scabrid   2
2. Leaves glabrous to softly pubescent or velvety; plants of
wetter mainly coastal areas   3
Leaves very rough with tubercle-based bristly white hairs;
plants of dry bushland   . . . .   4. *B. teitensis*
3. Leaves glabrous or almost so   . . . .   1. *B. petiolaris*
Leaves softly pubescent to velvety beneath   . . . . .   2. *B. nemoralis*

1. **B. petiolaris** (*Lam.*) *Thulin* in Nordic Journ. Bot. 7: 414 (1987); Verdc. in K.B. 43: 666 (1988); Martins in F.Z. 7(4): 73, t. 22 (1990). Type: shrub cultivated in Paris 'au Jardin du Roi' thought wrongly to have originated from material from the Antilles (no specimen found in Herb. Lam. which is clearly this but there is an unlabelled sheet)*

Shrub or tree 0.3–7.5(–12) m. tall, stated occasionally to be ± scandent; stems often hollow, pale brown, the older nodular with petiole-bases; epidermis white, peeling to reveal purple-brown beneath and dense lenticels; lower bark smooth, grey-brown or sometimes grey, rough and longitudinally fissured. Leaf-blades subcoriaceous, oblong-elliptic to obovate-elliptic, 1–14.5 cm. long, 0.7–7.5 cm. wide, rounded or ± acute at the apex, ± cuneate to rounded at the base, glabrous (see note) or almost so; petioles slender, 0.5–5.2 cm. long. Flowers sweetly scented. Cymes lax, ± pendulous, up to ± 15 cm. long; peduncles and secondary peduncles 1–2.5(–5) cm. long; pedicels 1–2 mm. long; axes ± pubescent. Calyx 3.5 mm. long, glabrous to pubescent outside, pubescent inside, divided into narrowly triangular lobes 1.5–2 mm. long, 1–2 mm. wide. Corolla waxy white, campanulate; tube 3–5 mm. long; lobes reflexed, broadly triangular or ovate with rounded apex, 0.5–1.5 mm. long. Anthers mauve. Style 2.5 mm. long, unbranched with ± capitate ± bifid green stigma or style occasionally stated to be very shortly bifid. Fruits orange-yellow to red, globose, ± 6 mm. diameter; pyrenes ¼-spherical, 4 mm. long, 2.8 mm. wide; outer curved face with close ridges bearing overlapping wings, the inner faces ± smooth.

KENYA. Kwale District: Tiwi Beach, 19 May 1979, *Bridson* 129!; Kilifi District; Kibarani, 16 Feb. 1945, *Jeffery* K75! & Vipingo, 17 Dec. 1953, *Verdcourt* 1081!
TANZANIA. Tanga District: Sawa, 13 Apr. 1955, *Faulkner* 1592!; Pangani District; Maziwi I., 1 Aug. 1974, *Frazier* 1041!; Uzaramo District: Wazo Hill, 5 Aug. 1967, *Harris* 732!; Pemba I., Ras Kigomasha, 10 Dec. 1930, *Greenway* 2686!
DISTR. **K** 7; **T** 3, 6, 8 (see note); **Z**; **P**; Mozambique, Madagascar and Mascarene Is.
HAB. Coastal dry lowland forest, e.g. *Cynometra, Manilkara, Afzelia*, littoral scrub, coral cliffs, also *Guettarda, Cordia subcordata*, etc., association just above high-tide mark and on sand-dunes; 0–30(–660) m. (see note), but mainly at sea-level
SYN. *Ehretia petiolaris* Lam., Encycl. Méth. Bot. 1: 527 (1785); Boj., Hort. Maurit.: 235 (1837); Bak., Fl. Maurit. & Seychelles: 201 (1877); Gürke in E. & P., Pf. IV. 3A: 87 (1893) & in P.O.A. C: 336 (1895); Bak. & Wright in F.T.A. 4(2): 21 (1905); R.C. Vaughan in Maurit. Inst. Bull. 1: 61 (1937); T.T.C.L.: 77 (1949); K.T.S.: 73 (1961)
    E. internodis L'Hérit., Stirp. Nov. 1: 47, t. 24 (1786), *nom. illegit.* Type as for *B. petiolaris*
    E. laxa Jacq., Hort. Schoenbr. 1: 18, t. 41 (1797); Boj., Hort. Maurit.: 236 (1837). Type: grown in Vienna from Réunion material (?W, holo.)
    Hilsenbergia ehretia Meissn., Pl. Vasc. Gen. Comment.: 198 (1840)**; Dunal in DC., Prodr. 13: 478 (1852). Type: Mauritius, *Sieber*, Fl. Maurit. 2: 160 (K, iso.!)
    H. rugosa Dunal in DC., Prodr. 13: 478 (1852). Type: Mauritius, *Sieber* 351 (G-DC, holo., microfiche!)
    Premna macrodonta Bak. in F.T.A. 5: 291 (1900). Type: Kenya, Mombasa, *Scott Elliot* 6106 (K, holo.!)

* The Commerson sheet (P-JU 6505) sometimes taken as the type is specifically said by Lamarck to be a variety and clearly should not be chosen.
  ** The earlier mention by Reichenbach (Consp. Regn. Veg.: 117 (1828)) is not accompanied by a description.

NOTE. *Sangster* s.n. (Lindi District, Rondo Plateau, Ngala, 660 m., Jan. 1954) not only comes from a high altitude for the species but is also said to be an epiphyte on *Brachylaena* ("no visible attachment to the ground"); it is, I believe, only a form. Some Mascarene specimens have irregular dense patches of indumentum on the undersides of the leaves which I suspect are of pathological origin.

2. **B. nemoralis** (*Gürke*) *Thulin* in Nordic Journ. Bot. 7: 415 (1987); Martins in F.Z. 7(4): 72 (1990). Types: Tanzania, Tanga District, Amboni, no collector mentioned, presumably *Holst* 2814 (B, syn.†, K, isosyn.!) & Bagamoyo, no collector mentioned (B, syn.†)*

Almost identical with the last, save for the velvety pubescence on the stems, undersides of the leaf-blades, inflorescence-axes, calyx, etc. and sometimes few to many additional glandular hairs. The leaf-blades can be rather larger, up to 18 cm. long, 8 cm. wide, sometimes acuminate at the apex, subcordate or cordate at the base but mostly cuneate to rounded. The inflorescences are also often rather more extensive with secondary peduncles up to 8 cm.

KENYA. Kwale District: Mwele Ndogo Forest, 4 Feb. 1953, *Drummond & Hemsley* 1106! & Marere area, 3 Apr. 1968, *Magogo & Glover* 727!; Lamu District: Kiunga, 16 Dec. 1946, *J. Adamson* 289!
TANZANIA. Tanga District: Steinbruch Forest, June 1951, *Eggeling* 6141!; Pangani District: Madanga, 2 Oct. 1951, *Tanner* 3148!; Bagamoyo District: Kikoka Forest Reserve, Apr. 1964, *Semsei* 3764!
DISTR. **K** 7; **T** 3, 6, 8; Mozambique
HAB. Lowland forest, woodland, bushland, thicket on coral, also in cultivations; 0–600 m.

SYN. *Ehretia litoralis* Gürke in P.O.A. C: 335 (1895) & in E.J. 28: 310 (1900); Bak. & Wright in F.T.A. 4(2): 21 (1905); T.T.C.L.: 77 (1949); K.T.S. 73 (1961); Vollesen in Opera Bot. 59: 75 (1980), *non B. litoralis* J.D. Smith (1898). Types: Tanzania, Tanga, *Holst* 2115 (B, syn.†, K, isosyn.!) & Tanga to Mkulumuzi [Mkulumusi] ferry, *Volkens* 198 (B, syn.†)
E. nemoralis Gürke in P.O.A. C: 336 (1895) & in E.J. 28: 310 (1900)

NOTE. This is probably no more than a variety of *B. petiolaris* but is distinctive with no intermediates; the distributions overlap but are not by any means the same, also *B. nemoralis* occurs more often at rather higher altitudes. A survey of the ecology of the two would help decide their true status.

3. **B. lyciacea** *Thulin* in Nordic Journ. Bot. 7: 415, figs. 1–3 (1987). Type: Somalia, Middle Juba region, 14 km. on Jelib–Camsuma road, near the village Shek Ahmed Yare, *Thulin & Warfa* 4506 (UPS, holo., K!, MOG, iso.)

Much-branched shrub 0.4–2.5(–4.5) m. tall, the upper branches often ± scandent; older stems pale grey, striate, with very reduced lateral branchlets bearing flowers and leaves, with very few bristly hairs on the young shoots but otherwise glabrous. Leaf-blades elliptic to obovate-elliptic, 0.5–3.3 cm. long, 0.35–1.8 cm. wide, rounded to truncate or emarginate at the apex, cuneate or attenuate at the base, ciliate with curved bristly hairs and often sparsely scabrid above with short tubercle-based hairs, glabrous beneath or with very few hairs; petiole 1.5–5 mm. long, bristly or glabrous. Flowers fragrant, solitary or sometimes 2(–3) from one cushion shoot or rarely in 2–3-flowered cymes, pendulous; pedicels 0.35–1.4 cm., lengthening to ± 2.5 cm. in fruit. Calyx cupular, the tube 1.5–2(–3) mm. long; lobes 5(–6), triangular, 1.5–3(–4) mm. long, 1.2–1.8 mm. wide, ciliate, tomentose inside. Corolla white or cream, turning pinkish buff or yellow or greenish white; tube 3 mm. long; lobes ovate, 4.5–5 mm. long, 3–3.8 mm. wide. Filaments ± 1 mm. long; anthers 1–2 mm. long. Ovary 7 mm. diameter; style ± 4 mm. long, bifid for 1.5–3 mm.; stigmas capitate or ± pileus-shaped. Fruits translucent orange, globose, 6–8 mm. diameter; endocarp bony, sharply and prominently ridged, breaking into 4 single-seeded pyrenes 4.8 × 3.2 × 2.8 mm.

KENYA. Northern Frontier Province: Dandu, 4 Apr. 1957, *Gillett* 12700! & Legumbisso, 28 Aug. 1945, *J. Adamson* 109!; Tana River District: Garissa–Garsen road, 8.3 km. towards Garsen from Bura turn-off, 8 July 1974, *R.B. & A.J. Faden* 74/1009!
DISTR. **K** 1, 7; Somalia, Ethiopia
HAB. *Commiphora-Grewia-Acacia* mixed bushland, *Terminalia orbicularis-Sesamothamnus* associations; 60–780 m.

SYN. *Ehretia sp. nov. aff. E. buxifolia* Willd. sensu K.T.S.: 73 (1961)

NOTE. An annotation by J.B. Gillett that two subspecies can be recognised on pedicel-length is not supported by additional material.

* In his second reference Gürke cites 9 *Stuhlmann* sheets from Bagamoyo and Dar es Salaam.

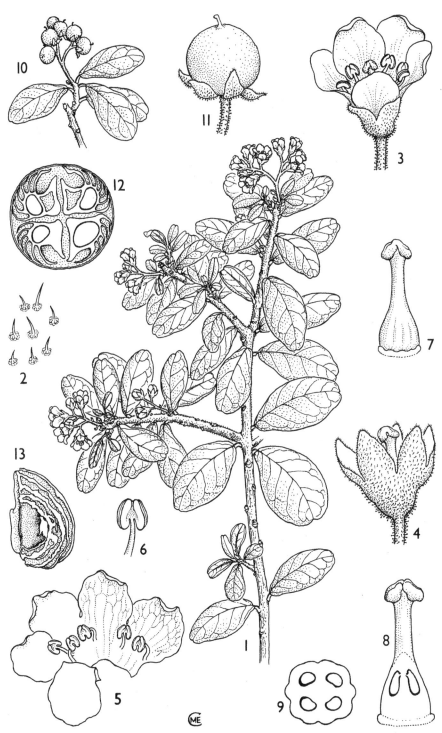

FIG. 10. *BOURRERIA TEITENSIS*—**1**, flowering branch, × ²/₃; **2**, upper leaf surface, × 16; **3**, flower, × 4; **4**, calyx, × 6; **5**, corolla, opened out, × 4; **6**, stamen, × 8; **7**, ovary, × 8; **8**, same, longitudinal section, × 10; **9**, same, transverse section, × 20; **10**, fruiting branchlet, × ²/₃; **11**, fruit, × 3; **12**, fruit, transverse section × 4; **13**, pyrene, × 6. 1–9, from *Drummond & Hemsley* 4267; 10, 11, from *Gilbert* 6113; 12, 13, from *Greenway* 6592. Drawn by Mrs Maureen Church.

4. B. teitensis (*Gürke*) *Thulin* in Nordic Journ. Bot. 7: 415 (1987). Types: Kenya, Teita, *Hildebrandt* 2359 & 2598 (B, syn.†) & NE. Tanzania, Nyika, *Holst* 2409 (B, syn.†, K, isosyn.!)

Shrub or small tree 1.8–6 m. tall with spreading twiggy branches; youngest shoots with spreading white pubescence, the older whitish grey, strongly wrinkled, glabrescent, lenticellate, roughened with raised petiole-scars. Leaf-blades brittle, oblong-elliptic to obovate-oblong, 0.5–5 cm. long, 0.3–3 cm. wide, rounded to emarginate at the apex, rounded, truncate, subcordate or ± cuneate at the base, coriaceous, the margins revolute, ± rugose or bullate with venation impressed above, discolorous, very rough above with tubercle-based stiff short white hairs and with some fine white pubescence above, glabrous beneath or with a few to ± dense white hairs on the venation. Cymes terminating the short shoots, short, ± few-flowered, with dense gland-tipped pale ferruginous pubescence on the axes; peduncles and secondary peduncles 0.5–2 cm. long; pedicels 2–6 mm. long. Calyx ± 7 mm. wide, deeply lobed, the lobes triangular or triangular-ovate, 2–3 mm. long, 2 mm. wide, pubescent outside like the axes. Corolla greenish yellow, campanulate; tube 3–4 mm. long; lobes ovate, 2.5–3 mm. long and wide. Anthers just exserted. Style 2 mm. long, very shortly bifid, the stigmas capitate. Fruit red, globose, 6 mm. diameter, brown and wrinkled when dry, with 4 pyrenes; weathered pyrenes with 5 distinct close wings on back. Fig. 10, p. 43.

KENYA. Kitui/Kilifi Districts: Lali Hills, Galana R., 30 Mar. 1963, *Thairu* 114!; Teita District: Mt.
   Kasigau, lower slopes above Rukanga, 4 Apr. 1969, *Faden et al.* 69/423!; Kwale District: Taru, 11
   Sept. 1953, *Drummond & Hemsley* 4267!
TANZANIA. Pare District: Buiko, 11 July 1942, *Greenway* 6592!; Lushoto District: Lake Manka, Apr.
   1966, *Procter* 3285! & Mkomazi valley, 30 Nov. 1935, *B.D. Burtt* 5316!
DISTR. **K** 4, 7; **T** 3; S. Somalia
HAB. *Acacia-Commiphora* bushland, *Terminalia orbicularis-Baphia-Grewia* thicket, also *Strychnos-*
   Euphorbia-Manilkara, etc. dry evergreen forest; 60–850 m.

SYN. *Ehretia teitensis* Gürke in P.O.A. C: 336 (1895); Bak. & Wright in F.T.A. 4(2): 21 (1905); T.T.C.L.:
   77 (1949); K.T.S.: 73 (1961)

NOTE. The record of this species from Ruwenzori (I.T.U., ed. 2: 50, footnote (1952)) is clearly
   erroneous and probably based on *Scott Elliot* 6172, in part, which bears a Ruwenzori Expedition
   label but is also clearly stated to be from British East Africa, Taru.

4. COLDENIA

L., Sp. Pl.: 125 (1753) & Gen. Pl., ed. 5: 61 (1754); I.M. Johnston in Journ. Arn. Arb. 32: 12
   (1951); A. Richardson in Sida 6: 235 (1976) & in Rhodora 79: 476 (1977)

Lobophyllum F.Muell. in Hook., Journ. Bot. 9: 21 (1857)

Procumbent herb with slender branched stems, often with adventitious roots. Leaves alternate, small, numerous, crenate, subsessile or petiolate. Flowers solitary, extra-axillary, 4-merous. Calyx deeply lobed. Corolla small; tube cylindrical, the throat naked ; lobes spreading. Stamens 4, inserted in the tube, included. Ovary 2-locular, each locule with 2 ovules or with 4, 1-ovuled locules; styles 2, terminal, partly joined at the base; stigmas scarcely differentiated. Fruits dry, ovoid-conic, 4-lobed, eventually dividing into 4, 1-seeded nutlets, joined amongst themselves or attached to a central prolongation of the receptacle. Seeds ovoid.

If considered in a wide sense, as has mostly been done for over 100 years, the genus has about 28 species all in the arid regions of America save for one species widespread in the Old World. Recently the New World species have been moved to *Tiquilia* Persoon by Richardson and I have followed the results of his careful and complete study. *Coldenia* is, therefore, here considered monotypic.

C. procumbens *L.*, Sp. Pl.: 125 (1753); C.B.Cl. in Fl. Brit. Ind. 4: 144 (1883); Hiern, Cat.
Afr. Pl. Welw. 1: 717 (1898); Bak. & Wright in F.T.A. 4(2): 28 (1905); I.M. Johnston in Journ.
Arn. Arb. 32: 13 (1951); F.P.S. 3: 77 (1956); E.P.A.: 771 (1961); Heine in F.W.T.A., ed. 2, 2:
321 (1963); Kazmi in Journ. Arn. Arb. 51: 148 (1970); Taton in F.C.B. Borag.: 26, t. 3 (1971);
Backer, Atlas 220 Weeds sugar-cane fields in Java, t. 513 (1973); Täckh., Students' Fl. Egypt,
ed. 2: 438 (1974); Richardson in Rhodora 79: 476, fig. 2, 3, 4* (1977); Vollesen in Opera

* This map is appallingly incomplete for Africa.

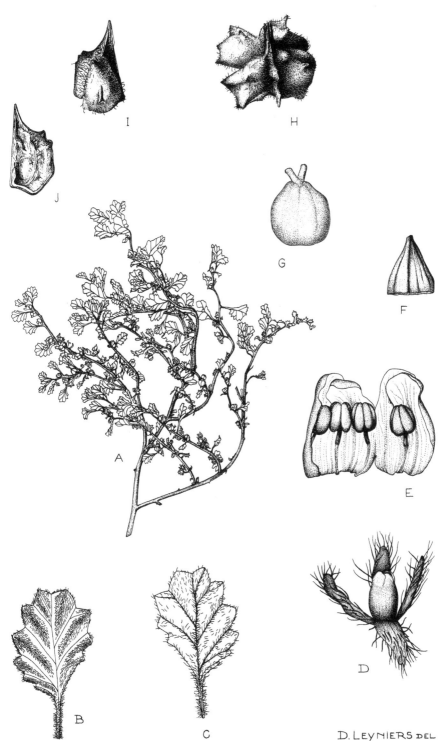

FIG. 11. *COLDENIA PROCUMBENS*—**A**, habit, × ½; **B**, leaf, upper surface, × 2 ½; **C**, same, undersurface, × 2 ½; **D**, young flower, × 12 ½; **E**, corolla, opened out, × 25; **F**, corolla, after flowering, × 8; **G**, ovary, × 25; **H**, fruit, × 7; **I**, nutlet, dorsal surface, × 7; **J**, same, ventral surface, × 7. All from *de Witte* 4880. Drawn by D. Leyniers. Reproduced with permission from Flore du Congo, du Rwanda et du Burundi.

Bot. 59: 75 (1980); Martins in F.Z. 7(4): 79, t. 24 (1990). Type: Ceylon, *Hermann Herbarium* (BM-HERM, lecto.!)*

Procumbent annual herb 10–50 cm. wide or forming mats 0.5–1 m. across; stems woody at the base, ± compressed, branched and ascending, greyish-velvety. Leaf-blades oblong, elliptic or obovate, asymmetric, 0.5–3.3 cm. long, 0.3–2.2 cm. wide, crenulate or lobulate, strigose between the nerves above with adpressed ± bulbous-based hairs pointing towards the apices of lobes making leaf appear plicate, with spreading hairs beneath; nerves in 4–6 pairs, impressed above; petiole 0.1–2 cm. long. Flowers small, subsessile. Calyx 2.5 mm. long, divided almost to the base into 4 lanceolate or ovate-lanceolate lobes, ± 0.6 mm. wide, often 1 lobe more developed, hairy, persistent. Corolla white, shorter than the calyx, glabrous, soon deciduous; tube ± cylindrical, 1.5–1.7 mm. long, slightly narrowed at the throat; lobes 4, oblong to obovate, 0.3–0.5 mm. long. Stamens 4; filaments 0.2–0.4 mm. long. Ovary conical, 0.4–0.6 mm. long, glandular-pubescent; styles 0.1–0.2 mm. long. Fruit 3–4 mm. long, glandular-hairy, apiculate, at first dividing into 2 then finally into 4 apiculate somewhat woody nutlets with rounded dorsal face with tubercles and irregular protuberances and angular ventral face. Seeds flattened-ovoid, 1.8 mm. long, 1.2 mm. wide. Fig. 11.

UGANDA. Toro District: Lake Albert, Toro Game Reserve, 15 Jan. 1971, *Wendelberger* U.86!

KENYA. Turkana District: Lodwar, Jan. 1932, *Champion* T16!; Lake Baringo, 12 Mar. 1950, *Bally* 7739!; Teita District: Tsavo National Park East, Aruba Dam at Voi R. inlet, 19 Jan. 1967, *Greenway & Kanuri* 13061!

TANZANIA. Pare District: Lake Kalimawe, 10 Jan. 1967, *Richards* 21928!; Rufiji District, Mafia I., Kirongwe [Kerongwe], 26 Aug. 1937, *Greenway* 5133!; Iringa District: Ruaha R., about 4 km. W. of Mtera bridge, 14 Aug. 1970, *Thulin & Mhoro* 730!; Zanzibar I., Sept. 1895, *Sacleux* 1036!

DISTR. U 2; K ?1, 2, 3, 7; T 3, 6–8; Z; Mauritania to Chad, Sudan and S. Egypt; Senegal to Angola, Ethiopia, Zaire, Mozambique, Zambia, Zimbabwe, Madagascar, tropical Asia and Australia

HAB. Dry mud of river banks and lake-shores, black cotton soil around ponds and rice cultivations near the coast, often with *Eclipta, Glinus, Cynodon,* etc., depressions and pans with standing water; 10–700 m.

SYN. *C. angolensis* Welw., Apont. Phytogeo.: 591 (1859). Type: Angola, Barra do Dande, near Bumbo, *Welwitsch* 5445 (LISU, holo., BM, iso.!)

     Lobophyllum tetrandum F.Muell. in Hook., Journ. Bot. 9: 21 (1857). Type: Australia, R. Victoria & Sturt's Creek, *F. Mueller* (MEL, syn. K, isosyn.!)

5. ARGUSIA

Boehmer in Ludw. Defin. Gen. Pl.: 507 (1760); Dandy, Ind. Gen. Vasc. Pl. 1753–74 (Regn. Veg. 51): 28 (1967) & in J.L.S. 65: 256 (1972)

Messerschmidia Hebenstr. in Nov. Comment. Acad. Sci. Imp. Petrop. 8: 315, t. 11 (1763); I.M. Johnston in Journ. Arn. Arb. 16: 161 (1935) (extensive references)**

Herbs, shrubs or trees, usually in saline or strand conditions. Leaves alternate, entire, attenuate at the base into a short or ± obsolete petiole, densely silky-hairy. Inflorescences dichotomously branched, the flowers in unilateral scorpioid cymes, dense or more lax, sessile or subsessile, white or greenish. Sepals 4–5, cuneate, round or oblong. Corolla subcylindrical or campanulate, the lobes imbricate or plicate in bud, later spreading. Stamens 4–5, included or partly exserted; anthers 2 to several times as long as wide; filaments very short. Ovary 4-locular with 1 ovule in each locule; style terminal, short, the stigma conical with basal receptive annular part and lobed apical part in the widespread species but in others it is conical, truncate-conic or bun-shaped. Fruits dry when mature, the carpels embedded in the centre of the corky exocarp or occupying the apical half of fruit, the lower half entirely 'spongy-corky' (see note); endocarp dividing into 2 parts, each with 2 seed-bearing cavities separated by a deep groove or sterile cavity.

Four species, one extending from Romania through central Asia to Korea, China and Japan, one in Turkmenistan, one in the West Indies and the other widespread in the Old World tropics especially on the shores of small islands. The fruits are well adapted for water dispersal.

     * Kazmi cites *Linn. Herb.* 174.1 as the type, but there is nothing to prove this specimen was there in 1753 and it is not annotated with specific name nor number.

     ** Name often misused for other groups of Boraginaceae. See Johnston, Contr. Gray Herb. 92: 73 (1930)

FIG. 12. *ARGUSIA ARGENTEA*—**1**, flowering branch, × ²⁄₅; **2**, infructescence, × ²⁄₅; **3**, flower bud, × 8; **4**, corolla, × 8; **5**, corolla, opened out, × 8; **6**, anther, × 10; **7**, gynoecium, × 8; **8**, fruit, × 4; **9**, same, transverse section (= 8a), × 4; **10**, same, longitudinal section (= 9b), × 4; **11**, same (= 9c), transverse section, × 4; **12**, seed, × 4. 1, 3–7, from *MacKee* 7900; 2, 8–12, from *MacKee* 19346. Drawn by G. Chypre. Reproduced with permission from Fl. Nouv.-Caléd.

A. argentea (*L.f.*) *Heine* in Fl. Nouv.-Caléd. 7: 109, t. 24, map 21 (1976); Martins in F.Z. 7(4): 81, t. 25 (1990). Type: Ceylon, *Koenig* (LINN 193/6, labelled "*argentifolia*", lecto.!)

Fleshy-leaved tree or shrub 1–10 m. tall with dome-shaped crown; branches and all other parts densely covered with greyish white ± silvery or golden silky hairs. Leaves densely placed; blades obovate, 10–30 cm. long, 3.5–11 cm. wide, rounded to obtuse or subacute at the apex, long-attenuate and decurrent into the 0.5–2.2 cm. long petiole; fallen leaves producing distinctive petiolar scars. Inflorescences terminal, at first ± globose, becoming paniculate-pyramidal, erect, 10–30 cm. long, 15–20 cm. wide; peduncle 5–15 cm. long, stout and dichotomous. Flowers sessile, very numerous, very fragrant, densely placed in two ranks on the 2–10 cm. long cymes. Calyx 1.5–2(–3) mm. long with slightly fleshy triangular-lanceolate to almost round lobes. Corolla white to pinkish; tube subcylindrical, 1.5–2 mm. long; limb 4–7 mm. wide, the lobes ovate or rounded, imbricate in bud. Anthers with tips just exserted. Ovary subglobose, 0.8–1.5 mm. diameter; stigmatic disc crowned by 2 rigid conic lobes. Fruit globose, 4–8 mm. diameter, the upper part occupied by the endocarp which breaks into 2 parts, each ± hemispherical with 2 locules containing the seeds and a small intermediary sterile locule; apex and dorsal surface of the carpel covered with spongy exocarpial tissue. Fig. 12, p. 47.

KENYA. Kilifi District: SE. of Gedi, 5 Jan. 1962, *Greenway* 10442!
TANZANIA. Zanzibar I., Kiwengwa, 29 Jan. 1929, *Greenway* 1242! & Chuaka, 27 Dec. 1930, *Vaughan* 1783!
DISTR. **K** 7; **Z**; Mozambique, Madagascar to tropical Asia, Malaysia, Vietnam, tropical Australia, Taiwan, Ryukyu Is. and Polynesia (Johnston gives detailed distribution)
HAB. Sea-shore just above high-water mark, sand-dunes with *Casuarina*, *Scaevola* etc.; ± sea-level

SYN. *Tournefortia argentea* L.f., Suppl. Pl.: 133 (1781); Klotzsch in Peters, Reise Mossamb., Bot.: 250 (1861); Bak. & Wright in F.T.A. 4(2): 29 (1905), excluding citation from Amboland; Verdc. in Kirkia 5: 271 (1966); Fosberg & Renvoize, Fl. Aldabra: 196, fig. 31/9–10 (1980)
     Messerschmidia argentea (L.f.) I.M. Johnston in Journ. Arn. Arb. 16: 164 (1935); Ju-Ying Hsiao in Fl. Taiwan 4: 404 (1978)

NOTE. Fosberg & Renvoize claim the fruit mesocarp lacks cork as is usually stated but is composed of very coarse firm aerogenous tissue. Examination at Kew by anatomists has confirmed that the thick mesocarp consists of rather large polyhedral parenchymatous cells, partially lignified but with no suberin.

6. **TOURNEFORTIA**

L., Sp. Pl.: 140 (1753) & Gen. Pl., ed. 5: 68 (1754); I.M. Johnston in Journ. Arn. Arb. 16: 145–161 (1935)

Shrubs or woody climbers with mostly broad petiolate leaves. Flowers in scorpioid cymes, spike-like, often arranged in dichotomous panicles. Calyx usually 5-lobed, persistent. Corolla mostly white or yellowish, 4–5-merous; tube cylindrical with spreading limb. Stamens usually 5, inserted in the tube, included, the filaments short. Ovary 4-locular; stigma sessile or borne on a distinct terminal style, peltate or conic, the stigmatic surfaces lateral, often bifid at the apex. Fruit a lobed or unlobed drupe, breaking up at maturity into 2–4 bony nutlets or pyrenes, each 1–2-seeded and often with 0–2 empty cavities; endosperm thin; cotyledons flat.

A genus of about 100 species, many variable and ill-defined, pantropical but particularly in America. Only a single poorly known species in East Africa.

T. usambarensis (*Verdc.*) *Verdc.* in K.B. 43: 436 (1988). Type: Tanzania, E. Usambara Mts., Mt. Bomole, *Zimmermann* in *Herb. Amani* 8058 (EA, holo.!)

Climbing shrub to 3 m. or more, or herb with rather slender brownish-pubescent stems. Leaf-blades oblong or elliptic-oblong, 4.5–13 cm. long, 1.2–7 cm. wide, acute, acuminate or rarely ± obtuse at the apex, cuneate to rounded at the base, sparsely pubescent and micro-scabridulous above, pubescent on the nerves beneath; petiole 0.5–2.2 cm. long, pubescent. Inflorescences terminal, dichotomously cymose, the branches scorpioid; primary branches 3.5–8 cm. long, secondary 0.7–3 cm. long, pubescent; true pedicels 0.25–1 mm. long. Calyx-lobes narrowly triangular, 1.8–2.5 mm. long, 0.6 mm. wide, acute, pubescent. Corolla white, salver-shaped; tube cylindrical, 8–8.5 mm. long, 0.8–1.3 mm.

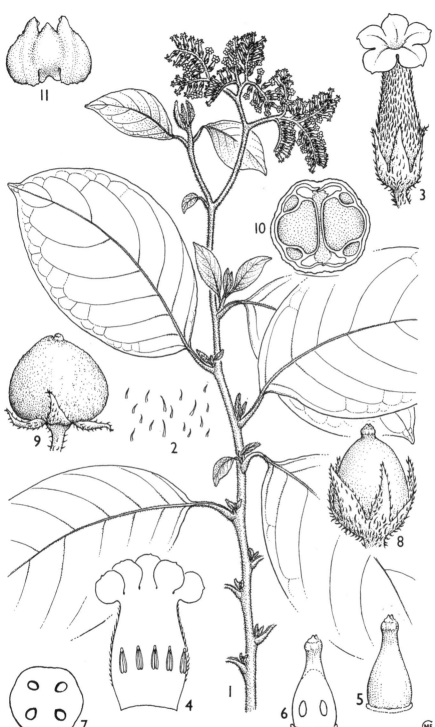

FIG. 13. *TOURNEFORTIA USAMBARENSIS*—**1**, habit, × ⅔; **2**, young leaf, upper surface, × 8; **3**, flower, × 8; **4**, corolla, opened out, × 8; **5**, ovary, × 16; **6**, same, longitudinal section, × 16; **7**, same, transverse section, × 30; **8**, young fruit, × 8; **9**, mature fruit, × 6; **10**, same, transverse section, × 6; **11**, nutlet, × 6. 1–7, from *Semsei* 3565; 8, from *Grote* 3582; 9, 10, from *Grote* 3937; 11, from *Iversen & Steiner* 86665. Drawn by Mrs Maureen Church.

wide, densely pubescent; limb 5 mm. wide, the lobes 1.5–1.7 mm. long, 2–2.2 mm. wide. Ovary conical, 0.8 mm. long, 0.6 mm. wide, truncate; style 1 mm. long, 0.5 mm. wide, the stigmatic part slightly thickened, 10-striate. Fruit ? ± fleshy, ovoid, 4 mm. long, 3.5 mm. wide, glabrous; exocarp ± fleshy, minutely rugulose; endocarp breaking into 2, 2-seeded nutlets, each with additional empty cavities. Fig. 13.

TANZANIA. Tanga District: E. Usambara Mts., Amani, Mt. Bomole, Nov. 1907, *Meyer in Herb. Amani* 1479! & Kwamkoro Forest Reserve, 27 Oct. 1962, *Semsei* 3565! & 6 km. NE. of Amani, 12 Nov. 1986, *Iverson & M. Steiner* 86665!; Morogoro District: E. Uluguru Mts., Tanana, 15 Feb. 1933, *Schlieben* 3453!

DISTR. **T** 3, 6; not known elsewhere

HAB. Rain-forest and derived cultivations; (?660–)900–1550 m.

SYN. *T. sarmentosa* Lam. subsp. *usambarensis* Verdc. in Kirkia 5: 272 (1966)

NOTE. When initially describing this I entirely overlooked Johnston's revision of the Old World species and the fact that *T. sarmentosa* is characterised by the fruit dividing eventually into 4 pyrenes. No material in European herbaria had been arranged by the revision. My original estimates of affinity with *T. viridiflora* Roth var. *griffithii* C.B.Cl. and *T. wightii* C.B.Cl. were accurate but both these are merged now with *T. montana* Lour. The rather different foliage and known difficulty of defining species throughout the range has suggested specific rank is advisable for the East African populations, but if not then it will be under *T. montana* that it must be placed at some rank. Obviously much commoner formerly; Peter collected 9 gatherings around Amani in four years whereas Greenway did not see it whilst living there over 20 years. Messrs Iversen & Steiner have collected it again recently.

7. HELIOTROPIUM

L., Sp. Pl.: 130 (1753) & Gen. Pl., ed. 5: 63 (1754)

Annual or perennial usually hairy herbs or subshrubs. Leaves alternate or in a few species opposite, sessile or petiolate, entire or denticulate. Cymes unilateral, spike-like, dense or lax, usually scorpioid, terminal, simple or branched, with or without bracts or flowers solitary at apices of leafy branches. Flowers small, ♂, regular. Calyx-lobes 5, lanceolate or linear, often slightly unequal, mostly shortly joined at base. Corolla white, yellow, cream or blue, tubular, salver-shaped or funnel-shaped, glabrous or pubescent, the throat often pubescent, usually without scales; lobes 5(–6), spreading, imbricate, folded and induplicate in bud, often with teeth between. Stamens 5(–6), included, the filaments very short or absent; anthers elliptic-oblong or lanceolate. Ovary completely or incompletely 4(–5)-locular with 1 ovule in each locule, often encircled by a basal disc; style terminal, short to long, or often absent; stigma depressed, conic or discoid, generally topped with a sterile cylindrical or conical appendage, entire or 3–4-lobed. Fruit dry or slightly fleshy, breaking into 4 distinct 1-seeded nutlets at maturity or cohering in pairs or reduced to 1 nutlet by abortion. Seeds straight or curved, generally with endosperm.

About 280 species in tropics to warm temperate regions, particularly in more or less arid areas. The genus has been divided into subgenera and a number of sections but no adequate complete modern infrageneric classification is available. Sectional names have sometimes been used in different ways by different authors and confusion has resulted. Riedl (Fl. Iranica, Boraginaceae (1967) and in Ann. Naturhist. Mus. Wien 69: 81–93 (1966)) and I.M. Johnston (Contrib. Gray Herb. 81 (1928)) have discussed the sections insofar as was necessary for treating the Iranian, European and S. American species. I have followed Riedl in accepting three subgenera.

Subgen. **Heliotropium**
Sect. *Heliotropium*, species 1, 2
Sect. *Gottliebia* Verdc.*, species 3
Sect. *Pterotropium* (DC.) Riedl (1967), *H. pterocarpum* (DC. & A.DC.) Bunge
Sect. *Coeloma* (DC.) I.M. Johnston, (1928), species 4–10
Sect. *Halmyrophila* I.M. Johnston, (1928), species 11
Sect. *Rutidotheca* (A.DC.) Verdc.** (sect. *Lophocarpa* Bunge, 1869), species 12

* Nom. nov. for Sect. *Messerschmidia* (DC.) Riedl, in Fl. Iranica 48: 16 (1967), who states it is based on "*Tournefortia* L. sect. *Messerschmidia* DC. *non Messerschmidia* Hebenstr." Type of section: *Heliotropium zeylanicum* (Burm.f.) Lam. Riedl uses the name for both a separate genus and a section of *Heliotropium*, but DC. refers back to *Messerschmidia* L., a very different plant from the three Riedl includes in the section. Since both usages are based on the same type the section is actually without a valid name.

** Comb. nov., *Heliophytum* DC. sect. *Rutidotheca* A.DC. in DC., Prodr. 9: 555 (1845).

Sect. *Heliothamnus* I.M. Johnston, (1928), *H. arborescens* L.
Sect. *Tiaridium* (Lehm.) Griseb. (1861), species 13

Subgen. **Orthostachys** (R.Br.) Riedl, (1967)
Sect. *Bracteata* (I.M. Johnston) Riedl (1967), species 14–19

Subgen. **Piptoclaina** (G. Don) Riedl, (1966)
Sect. *Piptoclaina*, species 20

Heliotropium arborescens L. (often still better known as *H. peruvianum* L. or *H. corymbosum* Ruiz & Pavon) the well known subshrubby 'cherry pie', with innumerable cultivars, and scented purplish flowers, is probably widely grown (see Jex-Blake, Gard. E. Afr., ed. 4: 115(1957)); specimens have been seen from Nairobi Arboretum (11 May 1953, *G.R. Williams Sangai* 546). It has been included in the key. A specimen of *H. pterocarpum* (DC. & A.DC.) Bunge at Kew labelled "Amani 1929", *Toms* 5 is puzzling. No true locality is appended to the specimen. Toms worked at Amani as caretaker/gardener before its revival as a research station. The specimen could have either been grown at Amani or more probably received from elsewhere for identification. It is very unlikely to grow wild in our area even in NE. Kenya. It is mentioned in the key.

In E. and NE. Africa the genus has been considered important since it is common in the preferred initial swarming areas of young hoppers of the desert locust although some species at least are stated not to be eaten by locusts, a fact noted by C.B. Williams at Naivasha in 1928.

1. Corolla-lobes drawn out into slender bent subulate tails
   (very common and widespread) . . . .   3. *H. zeylanicum*
   Corolla-lobes not drawn out or if so then appendages
   triangular or oblong, flat 2
2. Plant prostrate 3
   Plant erect or suberect (some species included twice if
   variable) 10
3. Corolla bright yellow; fruit characteristic, rostrate (in Flora
   area) . . . .   17. *H. baclei*
   Corolla not bright yellow, often white, sometimes cream or
   pale yellow-green 4
4. Glabrous subsucculent herb, ± glaucous when dry
   (adventive often near saline lake-shores) . . . .   11. *H. curassavicum*
   Not as above, usually densely pubescent 5
5. Leaves elliptic to ovate, often bullate-plicate and mostly
   silvery white hairy 6
   Leaves linear to narrowly elliptic, not so distinctly
   plicate 7
6. Leaves 1–11.5 × 1–5.5 cm.; corolla 3–6 mm. long; fruit
   breaking into 2, 2-seeded nutlets . . . .   8. *H. simile*
   Leaves 0.8–3.5 × 0.6–1.5 cm.; corolla 2.5 mm. long; fruit
   breaking into 1–4, 1-seeded nutlets . . . . .   20. *H. supinum*
7. Stigmatic club subglobose, the style obsolete or very short,
   often hidden between nutlets in fruiting state; foliage
   densely silvery strigose . . . .   16. *H. sessilistigma*
   Stigmatic club mitriform or conical with basal annular
   stigmatic area and narrower apex; style obsolete to
   well-developed 8
8. Plant of coastal dunes and nearby, straggling to erect;
   leaves elliptic, sparsely hairy, green; style as long as the
   conical stigmatic club . . . .   5. *H. gorinii*
   Not growing on coastal dunes; leaves silvery strigose;
   stigmatic club with prominent annulus and narrow
   apex (fig. 19/2, p. 71) 9
9. Leaves linear to linear-lanceolate, 0.4–3.6 cm. × 1–5 mm.;
   flowers less congested on spikes; style shorter than to
   almost equalling the stigmatic club . . . . .   15. *H. strigosum*
   Leaves obovate to elliptic, 0.5–5.5 × 0.25–2.5 cm.; flowers
   often very congested on spikes; style obsolete   19. *H. ovalifolium*
10. Subshrubby blue- or pale to deep purple-flowered plant
    with ± bullate leaves (cultivated) . . . .   *H. arborescens*
    Flowers not blue or purple 11

11. Nutlets 2, distinctly laterally winged; leaves elliptic, small,
    crenate; plant densely bristly white hairy (very doubtful
    if it actually occurs in Flora area — see note) . . .   *H. pterocarpum*
    Nutlets 2–4, if winged plant not bristly white hairy 12
12. Style scarcely or not developed, the stigmatic club
    practically sessile 13
    Style short to well-developed 14
13. Stigmatic club subglobose without apical hairs; leaves
    densely silvery white strigose; sepals subequal;
    intricately branched, suberect or prostrate subshrub   16. *H. sessilistigma*
    Stigmatic club conical, papillate, with apical tuft of hairs;
    leaves more sparsely strigose, up to 7 × 1.5 cm.; sepals ±
    unequal; more laxly branched herb of uncertain habit
    (**T** 4) .   .   .   .   18. *H. bullockii*
14. Leaves linear to linear-lanceolate, sessile or shortly
    petiolate, up to 5(–10) rarely 15 mm. wide 15
    Leaves oblong-elliptic or ovate to ovate-lanceolate, sessile
    to long-petiolate, mostly well over 15 mm. wide or if not
    then not linear 17
15. Branches with characteristic white peeling epidermis;
    stigmatic club ± globose, shorter than the very distinct
    style; shrub of ± arid areas   .   .   .   .   .   .   14. *H. rariflorum*
    Branches without peeling white epidermis; stigmatic club
    elongate 16
16. Style longer than stigmatic process which is apically bifid;
    leaves 1–8 × 0.1–1.5 cm., more sparingly strigose;
    nutlets rugose, nodulose or ± winged   .   .   .   .   12. *H. longiflorum*
    Style equalling or shorter than the stigmatic process which
    has a basal prominent annulus and narrower apex
    which is not bifid; leaves 0.4–3.6 × 0.1–0.5 cm., usually
    densely strigose   .   .   .   .   15. *H. strigosum*
17. Stems, leaves, calyx etc. softly velvety, not at all or scarcely
    harsh; nutlets 4, small, rugulose without lines or
    grooves 18
    Stems etc. with more strigose or at least usually ± harsh
    indumentum; nutlets 2 or if 4 then with ribs, grooved or
    smooth 19
18. Plant greener; corolla-tube 5 mm. long; style 2.5 mm. long
    (**K** 6; **T** 2)   .   .   .   .   2. *H. geissii*
    Plant more silvery; corolla-tube 3.3 mm. long; style 1 mm.
    long (**K** 1)   .   .   .   .   1. *H. aegyptiacum*
19. Nutlets 4, 1-seeded, ± beaked with fine raised ribs
    anastomosing at apex; annual lowland weedy herb
    0.2–1.5 m. tall with coarse white bristly indumentum;
    leaves 3–16 × 1.3–10 cm., ± bullate   .   .   .   .   .   13. *H. indicum*
    Nutlets 2, 2-seeded, without fine ribs but often sulcate or
    obtusely ribbed; leaves mostly smaller (save in
    H. scotteae) 20
20. Perennial upland ± forest plant 0.9–2.5 m. tall with large
    ± ovate leaves 3–15 × 2–10 cm. (**K** 1, ?3, 4)   .   .   .   7. *H. scotteae*
    Annual or perennial herbs of lower habitats with mostly
    much smaller leaves 21
21. Nutlets 2 with ± produced symmetrical apex of
    two diverging ± blunt angles with sinus between (see
    fig. 16/3d, p. 60) 22
    Nutlets 2, asymmetrical but sometimes with a curved
    groove ending in a ± lateral sinus (see fig. 16/4d, p. 60)
    or if symmetrical then ± rounded at apex and without
    obvious angles (fig. 14/1d, p. 54) 24

22. Leaves narrowly oblong-elliptic, up to 1 cm. wide, with petiole ± 5 mm. long; plant very sparsely hairy (**T** 3) — 6. *H. pectinatum* subsp. *mkomaziense*

Leaves ovate to elliptic or lanceolate, up to 4 cm. wide, with petiole to 2.5 cm. long; plant sparsely to densely hairy 23
23. Annual herb of coastal dunes with leaves 1–4.5 × 0.7–3.5 cm. — 4. *H. benadirense*

Perennial herb of inland ± arid areas with leaves to 9.5 × 4 cm.; petioles often very long in proportion to blades — 6. *H. pectinatum* subsp. *septentrionale*

24. Petioles scarcely developed or sometimes those of lower leaves up to 5 mm. long, rarely longer than $\frac{1}{15}$ of the blade; calyx-lobes mostly longer, 4–5 mm. long 25

Petioles mostly well-developed, (0.3–)1–2.5 cm. long and/or $\frac{1}{8}$–$\frac{4}{5}$ the length of the blade; calyx-lobes mostly shorter, 1–3 mm. long 26
25. Cymes unbranched or 2-branched . . . . — 9. *H. steudneri*

Cymes 3–5-branched . . . . — 10. *H. sp.* A
26. Plant of coastal sand dunes, straggling and matted, with mostly opposite subobtuse elliptic leaves 1–3.5 × 0.4–2.2 cm.; petiole 0.5–1.4 cm. long; inflorescences unbranched; corolla mostly glabrous or glabrescent outside . . . . — 5. *H. gorinii*

Plant of inland areas, unbranched annual or straggling perennial with opposite or alternate usually more acute or acuminate lanceolate to elliptic leaves 2–9.5 × 0.6–4 cm.; petiole up to 2.5 cm. long; inflorescences branched or unbranched; corolla often pubescent outside . . . . — 6. *H. pectinatum* (and variants) (note some forms of 8. *H. simile* will key near here)

1. **H. aegyptiacum** *Lehm.*, Ind. Sem. Hort. Hamb.: 20 (1824)*; F.P.S. 3: 82, fig. 18 (1956); E.P.A.: 771 (1961). Type: specimen cultivated in Hamburg from Egyptian seed, *Lehmann* (MEL?, holo., K, ? iso.)

Perennial herb to 75 cm. from a woody rootstock or rarely annual; stems branched, woody at base, softly velvety with dense grey spreading hairs which are not tubercle-based. Leaf-blades ovate to ovate-lanceolate, (1.5–)2–7(–11) cm. long, 0.8–3.5(–7.5) cm. wide, ± acute at the apex, broadly cuneate at the base, softly velvety like the stems; when worn the close granular-like tubercle-bases are visible; petioles 0.3–2(–4.5) cm. long. Inflorescence a panicle of scorpioid cymes, each eventually 4–10 cm. long, softly velvety; flowers numerous, densely placed, sessile. Calyx with indumentum similar to stems, 2.5–4.2 mm. long, the lobes linear-oblong to lanceolate, 2.2–4 mm. long, 0.5–0.8 mm. wide. Corolla white; tube narrow, slightly swollen below the middle, equalling or longer than the calyx, 3–4 mm. long, adpressed pubescent outside; limb ± 4–5 mm. wide, the lobes ovate or oblong, 2 mm. long, 1.5 mm. wide, crinkly, mostly with no interlobal teeth, adpressed pubescent on midrib outside. Anthers included or tips just exserted, 1.5 mm. long. Style stout, 0.6–1 mm. long, puberulous; stigma elongate-conic, 1.2 mm. long, puberulous, slightly bifid at apex, with lobulate ring at base but not produced. Nutlets ± trigonous, 1.8 mm. long, 0.8 mm. wide, slightly rugulose, very slightly winged at margins. Fig. 14/1, p.54.

KENYA. Northern Frontier Province: El Wak, 25 May 1952, *Gillett* 13336! & between Isiolo and Archer's Post, Uaso Nyiro valley at Buffalo Springs, 25 Jan. 1961, *Polhill* 331! & 4 km. S. of Archer's Post, 14 June 1979, *Gilbert et al.* 5658!
DISTR. **K** 1; Arabia, Egypt, Sudan, Ethiopia, Djibouti, Somalia
HAB. Overgrazed *Commiphora-Acacia* open scrub with succulents, also in *Acacia-Hyphaene* with *Barleria, Vernonia, Amaranthaceae* etc.; 360–770 m.

* The Index Kewensis gives page 8 (1820), but I have been unable to find a seed list dated 1820.

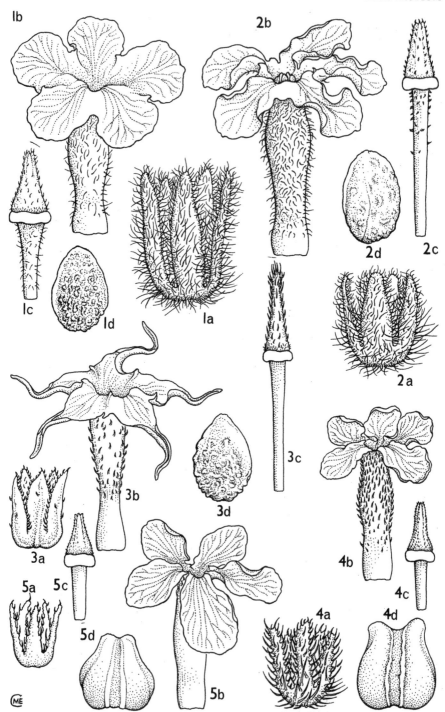

FIG. 14. *HELIOTROPIUM AEGYPTIACUM*—**1a**, calyx, × 10; **1b**, corolla, × 10; **1c**, style and stigma, × 20; **1d**, nutlet, × 12. *H. GEISSII*—**2a**, calyx, × 10; **2b**, corolla, × 10; **2c**, style and stigma, × 20; **2d**, nutlet, × 12. *H. ZEYLANICUM*—**3a**, calyx, × 10; **3b**, corolla, × 10; **3c**, style and stigma, × 20; **3d**, nutlet, × 12. *H. BENADIRENSE*—**4a**, calyx, × 10; **4b**, corolla, × 10; **4c**, style and stigma, × 20; **4d**, nutlet, × 12. *H. GORINII*—**5a**, calyx, × 10; **5b**, corolla, × 10; **5c**, style and stigma, × 20; **5d**, nutlet, × 12. 1, from *Hemming* 1086; 2a–c, from *Verdcourt* 3263; 2d, from *Nattrass* 1255; 3a–c, from *Greenway* 815; 3d, from *Faden & Ngweno* 74/817; 4, from *Gillespie* 85; 5, from *Greenway* 13138. Drawn by Mrs Maureen Church

SYN. *H. pallens* Del., Cent. Pl. Afr. Caill.: 69, t. 3/4 (1826) (preprint) & in Cailliaud, Voy. Méroé 4: 362 (1827) & Atlas (folio) t. 64/4 (1827); DC. & A.DC. in DC., Prodr. 9: 534 (1845); Schweinf., Beitr. Fl. Aethiop.: 116 (1867); Vatke in Oest. Bot. Zeitschr. 25: 166 (1875); Engl., Hochgebirgsfl. Trop. Afr.: 351 (1892); Bak. & Wright in F.T.A. 4(2): 33 (1905); Täckh., Students' Fl. Egypt, ed. 2: 442 (1974). Type: Sudan, Dongolah, *Cailliaud* (?MPU, holo.)

    H. cinerascens DC. & A.DC. in DC., Prodr. 9: 534 (1845); A.Rich., Tent. Fl. Abyss. 2: 85 (1850); Engl., Hochgebirgsfl. Trop. Afr.: 352 (1892); Bak. & Wright in F.T.A. 4(2): 39 (1905); E.P.A.: 772 (1961); Täckh., Students' Fl. Egypt, ed. 2: 441 (1974). Type: Ethiopia, without exact locality, *Schimper* 1161 (G, holo., BM, K, iso.!)

    [*H. europaeum* sensu Grant* in Oliv., App. Speke's Journ.: 641 (1863) & in Trans. Linn. Soc., Bot., 29: 114 (1875) *non* L.]

NOTE. For many years various routine namers have suggested in annotations that *H. pallens* and *H. cinerascens* were basically conspecific despite the wide divergence of the actual types. *H. pallens* is a woolly annual with broad leaves and *H. cinerascens* a perennial plant with smaller narrow leaves, more adpressed indumentum and woody branches. Most of the considerable material available is without a rootstock. The Kenya specimens have the indumentum of *pallens* but the perennial habit of *cinerascens*. There is too much overlap to keep them separate. Unfortunately *H. aegyptiacum* is older than either of these two well-known names and is the correct name for the species. *Herlocker* 451 (Northern Frontier Province, Marsabit District, edge of the Chalbi Desert, Kalacha, 9 Mar. 1977) has thick obtuse leaves but is probably only a variant of this species.

2. **H. geissii** *Friedrich* in Mitt. Bot. Staats. München 3: 616 (1960); Martins in F.Z. 7(4): 89, t. 26/5 (1990). Types: Namibia, Damaraland, Kalkfeld, *Dinter* 7486 (M, syn., K, isosyn.!), Wittklipp, *Volk* 2879 (M, syn.) & near track to Outjo, Omatjenne, *Volk* 2942 (M, syn.)

Perennial or probably occasionally annual erect ± branched herb (0.2–)0.4–1 m. tall, often in clumps; stems with dense spreading tubercle-based hairs and much shorter almost granular indumentum, probably somewhat glandular; rootstock creeping (*fide* U.K.W.F.). Leaf-blades drying a characteristic pale yellowish grey-green colour, elliptic-lanceolate, (2–)3.5–12 cm. long, (0.5–)1–3.5 cm. wide, narrowly rounded to ± acute at the apex, cuneate at the base, harshly velvety with tubercle-based hairs, particularly dense on the venation beneath, the yellowish tubercles appearing as if separate glands under a hand-lens but high magnification showing each one to bear a spreading hair; petiole 0.3–1 cm. long. Inflorescence a panicle of scorpioid cymes, each subglobose when young, 1–1.5 cm. wide but rapidly elongating up to 6 cm. long; flowers sessile, slightly scented. Calyx 3.5–4(–5) mm. long; lobes lanceolate, almost free, spreading pubescent. Corolla white or yellowish; tube 5–5.5(–6.5) mm. long, swollen in upper part in dried state, adpressed shortly pubescent outside; lobes oblong to obovate, 1.5–2.5 mm. long, 1–1.3 mm. wide, crinkly, rounded, sometimes with short interlobular narrowly triangular inflexed teeth. Anthers very narrowly sagittate, 1.7 mm. long, the apices usually just protruding. Style linear-obtriangular, 2.5–3 mm. long, with few retrorse hairs at apex or upper ⅔ quite distinctly retrorse pubescent, the green stigma conical, 1 mm. long, puberulous, slightly bifid and with lobulate ring at the base. Nutlets 3–4, cohering, 2.2 mm. long, 1.5 mm. wide, distinctly rugulose, but otherwise glabrous. Fig. 14/2.

KENYA. Nairobi District: Nairobi Royal National Park, Athi Gate, 24 Jan. 1962, *Verdcourt* 3263!; Masai District: SE. Ngong Hills, Magadi road, 2 July 1950, *Nattrass* in A.D. 3506! & ± 64 km. from Magadi on Nairobi road, 11 Sept. 1963, *Verdcourt* 3759!

TANZANIA. Masai District: between Longido and Engare Naibor, 25 Mar. 1970, *Richards* 25676! & 159 km. from Athi R. towards Arusha, 21 Sept. 1964, *Leippert* 5018! & foot of Mt. Longido, 25 July 1951, *Waloff* 2!

DISTR. **K** 3, 4, 6; **T** 2; Zimbabwe, Botswana, South Africa (Transvaal) and Namibia

HAB. Grassland, dry *Acacia* bushland, often by roadsides and on rendzinic soils; 1220–2070 m.

SYN. *H. erectum* M. Holz. in Mitt. Bot. Staats. München 1: 338 (1953), *non* Lam. (1778), *nec* Vell. (1825). Type as for *H. geissii*

    H. sp. near *H. cinerascens* sensu Verdc. in Heriz-Smith, Wild Fl. Nairobi Roy. Nat. Park: 53 (1962)

    H. sp. A sensu Kabuye & Agnew in U.K.W.F.: 520 (1974)

NOTE. A specimen from Nanyuki, Kenya, collected in 1926 by Mrs Prescott Decie is the most northerly known and also the first collected in East Africa.

---

* Oliver specifically states 'Col Grant's specimen I have not seen' — yet it is at Kew.

3. **H. zeylanicum** (*Burm.f.*) *Lam.*, Encycl. Méth. Bot. 3: 94 (1789); Wight, Ic. Pl. Ind. Or.
3: 8, t. 892 (1843); C.B.Cl. in Fl. Brit. Ind. 4: 148 (1883); Gürke in P.O.A. C: 336 (1895);
Hiern, Cat. Afr. Pl. Welw. 1: 720 (1898); Bak. & Wright in F.T.A. 4(2): 31 (1905); F.W.T.A. 2:
199 (1931); De Wild. & Staner, Contr. Fl. Katanga, Suppl. 4: 86 (1932); F.P.S. 3: 82 (1956);
E.P.A.: 778 (1962); Verdc. in K.B. 44: 166 (1989); Martins in F.Z. 7(4): 89, t. 26/4 (1990).
Type: India, Madras, Tutikorin, *Garcin* (G, holo.!)

Perennial suffruticose herb 25–80 cm. tall, erect or sometimes procumbent at the base
with a long thick rootstock or sometimes flowering ?as an annual or in its first year; stems
branched, woody at the base, sparsely to densely covered with tubercle-based hairs.
Leaf-blades linear to elliptic or sometimes lanceolate, 0.8–9(–12) cm. long, 0.2–1.2(–1.8 or
exceptionally –2.4) cm. wide, acute at the apex, narrowly attenuate at the base, entire,
sparsely to densely pubescent with tubercle-based hairs above and pubescent on
nervation beneath; petiole obsolete or very short. Cymes spike-like, one-sided, single or
paired, 1.5–30 cm. long; flowers sessile; bracts absent. Calyx lobed to the base, the lobes
narrowly elliptic, 1.5–2.2(–3 in fruit) mm. long, 0.4–0.9 mm. wide, thick, glabrous or
pubescent with margins ± undulate, sparsely ciliate, sparsely pubescent toward top inside.
Corolla with white tube and yellow-green limb extending along midrib into throat, or with
yellow-brown centre; tube ± cylindrical, slender, 2.8–4.5 mm. long, widened above,
pubescent outside in bands corresponding to the lobes; limb spreading, 4–8 mm.
(including tails) or ± 2 mm. (excluding tails) wide, the lobes triangular at base, 0.5–1 mm.
(excluding tails), caudate-acuminate, the tails subulate or flattened at base, 2–3.5(–4) mm.
long, pubescent outside. Style 0.7–1.5 mm. long, with a conic stigma 1.3–1.7 mm. long
terminated by a tuft of hairs. Fruit subglobose, 1.5 mm. tall, 2–2.6 mm. diameter, glabrous,
verrucose, breaking into 4 nutlets. Fig. 14/3, p. 54; 15.

UGANDA. Karamoja District: Mt. Kadam [Debasien], Jan. 1936, *Eggeling* 2552!; Teso District: Usuku,
Toroma, Nov. 1931, *Hansford* 2359!; Busoga District: Lake Victoria, Lugala Landing, 26 Mar. 1953,
G.H.S. Wood 654!
KENYA. Northern Frontier Province: Isiolo–Wajir road, 16 km. W. of Mado [Modo] Gashi, 8 Dec.
1977, *Stannard & Gilbert* 850!; Machakos District: Mtito Andei, Jan. 1950, *Bally* 7702!; Teita District:
near Tsavo railway station, 18 Jan. 1972, *Gillett* 19595!
TANZANIA. Masai/Mbulu Districts: top of Mto wa Mbu scarp, 7 Aug. 1956, *Verdcourt & Napper* 1564!;
about 8 km. S. of Dodoma, 20 July 1956, *Milne-Redhead & Taylor* 11184!; Kilosa District: Great
Ruaha valley, 20 km. SW. of Mikumi, 19 Mar. 1975, *Hooper & Townsend* 904!
DISTR. U 1, 3; **K** 1, 2, 4, 6, 7; **T** 1–8; Senegal, Mali, Niger, Nigeria, Cameroon, Zaire, Chad, Sudan,
Ethiopia, Somalia, Zambia, Zimbabwe, Malawi, Mozambique, Angola, Botswana, South Africa,
Arabia, Socotra, Comoro Is., India (?not in Ceylon)
HAB. *Acacia–Commiphora–Salvadora* etc. bushland, grassland with or without scattered trees, also as
a weed of roadsides and cultivations; 0–1650(–?1740) m.
SYN. *H. curassavicum* L. var. *zeylanicum* Burm.f., Fl. Indica : 41, t. 16/2 (1768), see note
    Tournefortia subulata A.DC., in DC., Prodr. 9: 528 (1845). Types: Senegal, *Perrottet* (G, syn., P,
      isosyn.) & Kouma, *Heudelot* (G, syn., P, isosyn.) & Ethiopia, Semien, Sabra, *Schimper* 1285 (G,
      syn., BM, K, isosyn.!) & Sudan, Kordofan, *Kotschy* 163 (G, syn., BM, K, isosyn.!)
    T. zeylanica (Burm.f.) Wight, Ill. Ind. Bot. 2: 211, t. 170/B (1850)
    T. stenoraca Klotzsch in Peters, Reise Mossamb., Bot.: 250 (1861). Type: Mozambique, Rios de
      Sena, *Peters* (B, holo.†, EA, iso.!)
    Heliotropium subulatum (A.DC.) Vatke in Linnaea 43: 316 (1882); Martelli, Fl. Bogos.: 59 (1886);
      Duthie, Fl. Gangetic Plain 2: 91 (1911); Heine in F.W.T.A., ed. 2, 2: 322 (1963); Kazmi in Journ.
      Arn. Arb. 51: 157 (1970); Taton in F.C.B., Borag.: 30, t. 4, map 9 (1971); Kabuye & Agnew in
      U.K.W.F.: 520, fig. 519 (1974); Vollesen in Opera Bot. 59: 76 (1980)
    H. zeylanicum (Burm.f.) Lam. var. *subulatum* (A.DC.) Chiov. in Fl. Somala 3: 140 (1936)
NOTE. The figure given by Burman is as Duthie (1911) and Trimen (Handb. Fl. Ceylon 3: 199,
adnot. (1895)) point out, either not the plant in question or a very poor representation of it since
the tails have been omitted from the corolla. I have, however, examined a Garcin specimen which
I am sure is the holotype and this is undoubtedly the plant which has long been known as
zeylanicum. The plant is commonly attacked by a gall which produces clusters of flask-shaped
densely pilose galled flowers 7 × 5 mm. Kazmi (Journ. Arn. Arb. 51: 155 (1970)) and many others
have considered the name *H. zeylanicum* to apply to a different species, *H. paniculatum* R.Br., but
did not examine the type. Chiovenda (Fl. Somala 3: 141 (1936)) recognises a variety *arenarium*
(Vatke) Chiov. (*Heliotropium subulatum* var. *arenarium* Vatke in Linnaea 43: 316 (1882); type:
Somalia, Brava, *Hildebrandt* 1315 (B, holo.†), see E.P.A.: 778 (1962)). It is a bristly pilose dune form.

4. **H. benadirense** *Chiov.*, Result. Sci. Miss. Stef.-Paoli, Coll. Bot.: 119 (1916) & Fl.
Somala 2: 319, fig. 185 (1932); E.P.A.: 772 (1961). Types: S. Somalia, between Mogadishu
and Gezira, *Paoli* 63, 110, 119 & Ras Deg-Deg, *Paoli* 294 (all FT, syn.)

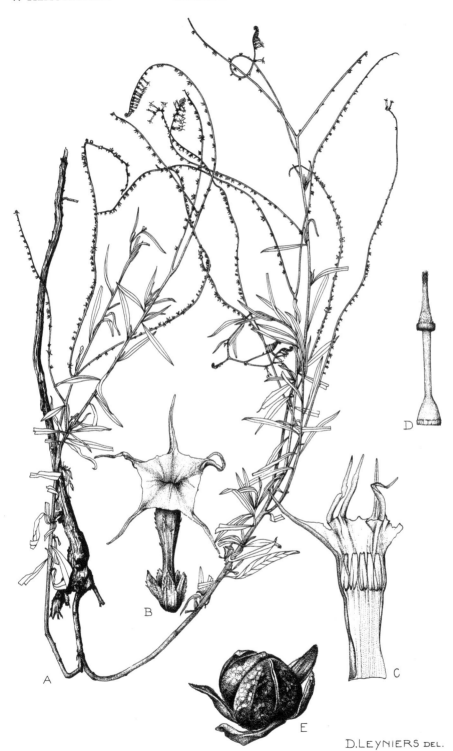

FIG. 15. *HELIOTROPIUM ZEYLANICUM*—**A**, habit, × ½; **B**, flower, × 8; **C**, corolla, opened out, × 8; **D**, gynoecium, × 12 ½; **E**, fruit, × 10. All from *Germain* 5507. Drawn by D. Leyniers. Reproduced with permission from Flore du Congo du Rwanda et du Burundi.

Annual branched slightly fleshy herb (3–)15–25(–30) cm. tall; stems with upwardly directed tubercle-based bristly white hairs but not obscuring the surface; rootstock slender, little-branched. Leaf-blades ovate, 1–3.5(–4.5) cm. long, 0.7–2(–3.5) cm. wide, ± acute to narrowly rounded at the apex, abruptly contracted to a cuneate base, with regular adpressed bristly white ± tubercle-based hairs above of unequal size and similar beneath with longer hairs on the nerves; petioles 0.3–1.2 cm. long. Scorpioid cymes both lateral and terminal, simple in the limited material seen, 1–9 cm. long, the flowers sessile, very dense and touching or at least closely placed, the axes with similar indumentum to the stems. Calyx ± 2.5–3 mm. long with white bristly spreading hairs; lobes linear-triangular, ± 0.5–1 mm. wide at the base. Corolla white; tube just under 3(–4 fide Chiovenda) mm. long, the upper ⅔ pubescent outside, the base glabrous; lobes oblong or ± round, 1.2–2 mm. long, 0.8–1.5 mm. wide, rounded. Anthers 0.9 mm. long, attached about the middle of the tube. Style 0.8 mm. long, shining, minutely striate; stigma 0.5 mm. long, the narrow apical part papillate, the wider coroniform part grooved. Fruit depressed ovoid-oblong, 2.5 mm. wide, 1.8 mm. tall, greenish brown, glabrous but clearly papillate under very high magnifications; nutlets 2, the sutures produced at apex laterally so that if viewed from side they appear shortly 2-horned, with 2 obscure furrows, each nutlet 2-seeded. Fig. 14/4, p. 54.

KENYA.   Lamu District: Kiunga, 24 July 1961, *Gillespie* 42! & 29 July 1961, *Gillespie* 85! & Mkokoni, Sept. 1956, *Rawlins* 154!
DISTR.   **K** 7; Somalia (S.)
HAB.   Sand-dunes and sandy places a little inland; 3–6 m.
NOTE.   *Jeffery* 209! (Kenya, Kilifi, 28 May 1945) is a large-leaved form.

5. **H. gorinii** *Chiov.* in Fl. Somala 2: 317, fig. 184 (1932); E.P.A.: 774 (1962). Type: S. Somalia, Transjuba, Kismayu, near El Ualud, *Gorini* 492 (FT, holo.)

Herb with erect, straggling or prostrate sparsely to much-branched stems forming tangled clumps 0.9–1.5 m. across and 15 cm. tall from a long slender annual (or ?sometimes perennial) root; stems scabrid with very few to sparse upwardly directed bristly tubercle-based hairs. Leaves ± subsucculent, mostly opposite, usually yellowish green, elliptic to ovate-elliptic, 1–3.5(–5.5) cm. long, 0.4–2.2 cm. wide, narrowly rounded to 'bluntly acute' at the apex, cuneate into the 0.5–1.4 cm. long petiole, with rather scattered to rarely ± dense short very broadly based hairs from cystolith spots, sometimes restricted to margins and main nerves. Flowers in long slender spike-like unbranched scorpioid cymes 7–23 cm. long; peduncle 2.5–5 cm. long, glabrous or with scattered hairs. Calyx 1.5–2 mm. long; lobes linear-lanceolate, glabrous or with scattered hairs. Corolla with white limb and greenish yellow tube; tube 3–4.5 mm. long, swollen in middle in dry state, glabrous or with scattered hairs; lobes ± unequal, obovate-spathulate, 1.5–2.5 mm. long and wide. Stamens placed just above middle of tube, the anthers 0.7 mm. long. Style 0.8 mm. long; stigma conical, 0.5 mm. long. Nutlets obliquely ovoid, 2–2.5 mm. long, 2 mm. wide with distinct median groove bounded by raised areas on each side then outer areas lower with slight groove between; main groove often asymmetric at tip of nutlet; 2 main plus 3 sterile loculi in section. Fig. 14/5, p.54.

KENYA.   Kwale District: Twiga, 13 Jan. 1964, *Verdcourt* 3909!; Mombasa District: Nyali Beach, 26 Apr. 1950, *Rayner* 296a!; Kilifi District: S. end of Shanzu Beach, 27 Dec. 1967, *Greenway* 13138!
DISTR.   **K** 7; S. Somalia
HAB.   Above high-tide in typical littoral associations of *Cocos*, *Hyphaene*, *Guettarda*, *Triainolepis*, *Scaevola*, *Hibiscus tiliaceus*, etc. but also on coral hillocks by shore and occasionally by roadsides and in cultivations; ± sea-level–15 m.

SYN.   [*H. longiflorum* sensu Vatke in Linnaea 43: 317 (1882) & Bak. & Wright in F.T.A. 4(2): 41 (1905), pro parte, *non* (A.DC.) Jaub. & Spach]

NOTE.   The type locality is some distance from the coast but still in an area of sandy dunes; the type has not been seen but the description agrees well with the plants cited above although Chiovenda mentions minute patent hairs on the corolla. *Ciferri* 51 (Somalia, Uebi Scebeli, Vittoria d'Africa, 17 Oct. 1934 (K!)) determined as *H. gorinii* by Chiovenda has more acute leaves, pubescent inflorescence calyx and corolla-tube and is closer to *H. pectinatum*. *H. gorinii* could easily be considered a subspecies of *H. pectinatum* but I have been influenced by the distinctive habit and ecology.

6. **H. pectinatum** *F. Vaupel* in E.J. 48: 530 (1912). Type: Kenya, Tana River District, Witu, Massa Malakofi, *F. Thomas* 15 (B, holo.†, A, fragment!)

Erect unbranched annual or shrubby often straggling perennial with woody stems and spreading lateral branches 10–60 cm. tall and up to 1 m. wide; stems with very sparse to fairly dense spreading or ± adpressed bristly white hairs. Leaves lanceolate or elliptic-lanceolate to elliptic-ovate, 2–9.5 cm. long, 0.6–4 cm. wide, usually narrowly acute at the apex but ± rounded in some variants, very gradually attenuate into the 0.3–2.5 cm. long petiole at the base, sometimes slightly crenulate, bristly pubescent on both sides with rather short subappressed stiff bulbous-based hairs and sometimes granular-tuberculate. Inflorescences bifid or unbranched, less often 3-fid, the cymes at first short and scorpioid, later long and spike-like, 3–19 cm. long; peduncles up to 15 cm. long; flowers sessile. Calyx-lobes linear-lanceolate, (1–)2–3(–5) mm. long, usually ± widened at base, up to ± 0.5 mm. wide, $\frac{1}{4}$–$\frac{2}{3}$ the length of the corolla-tube, very variable, bristly pubescent to ± glabrous. Corolla white, the centre mostly yellowish and lobes sometimes with dark brown marginal line; tube 2–7 mm. long, swollen below middle, mostly with few adpressed hairs at apex or sparsely bristly pubescent; lobes ovate or oblong, 1–3 mm. long, 1–2.5 mm. wide. Style ± 1–1.2 mm. long, the stigmatic cone 0.8–1 mm. long, narrowed above. Fruit depressed ovoid, ± 2–2.2 mm. long, 3.2–4 mm. wide, with 2 nutlets; nutlets with curvilinear depressed area near suture which forms an oblique sinus where it meets apex of nutlet but apex scarcely 2-horned or in some variants distinctly symmetrically 2-horned.

KEY TO INFRASPECIFIC VARIANTS

1. Leaves smaller, mostly under 5 × 2–3 cm. 2
   Leaves larger, up to 9.5 × 4 cm. 3
2. Leaves mostly over 1 cm. wide; fruits with nutlets having
   asymmetrical sinus at top (**K** 1, 4, 7, **T** 3) . . . .   a. subsp. **pectinatum**
   Leaves ± 7 mm. wide; nutlets with symmetrical sinus
   appearing distinctly 2-horned at apex (**T** 3) . . .   subsp. **mkomaziense**
3. Corolla-tube 2–2.5 mm. long; unbranched or sparsely
   branched annual; nutlets with asymmetrical sinus
   (**T** ?1, 4, 5, 7) . . . .   b. subsp. **harareense**
   Corolla-tube 3–7 mm. long; annual or perennial,
   sometimes spreading straggler 1 m. across; nutlets with
   asymmetrical or symmetrical sinus, sometimes 2-
   horned at apex . . . .   c. subsp. **septentrionale**

a. subsp. **pectinatum**

Erect unbranched or sparsely branched annual or straggling much-branched perennial with woody lower stems, 0.4–1 m. long or tall. Leaves ovate-lanceolate or lanceolate to ± 5 cm. long, 0.5–2(–3) cm. wide, usually acute or acuminate at the apex, cuneate at base into petiole up to 2.5 cm. long. Calyx-lobes (1–)2–3 mm. long. Corolla-tube 2.5–5 mm. long. Fruits with asymmetrical sinus at apex of nutlets. Fig. 16/1, p. 60.

KENYA. Northern Frontier Province; Isiolo–Wajir, 16 km. W. of Mado Gashi [Modo Gash], 8 Dec. 1977, *Stannard & Gilbert* 842!; Kitui District: Nairobi–Garissa road, 5 km. E. of Ukazzi, 9 May 1974, *Gillett & Gachathi* 20503!; Tana River District: Kurawa, 5 Oct. 1961, *Polhill & Paulo* 593!
TANZANIA. Lushoto District: Korogwe, Magunga Estate, 10 Oct. 1952, *Faulkner* 1056! & Mombo, 27 Dec. 1912, *Grote in Herb. Amani* 3936!; Pangani District: Between Hale and the Little Pangani Falls, 14 Aug. 1918, *Peter* 52447!; Zanzibar I., Mtoni, Feb. 1930, *Vaughan* 1290!
DISTR. **K** 1, 4, 7; **T** 3; **Z**; S. Somalia
HAB. Grassland, *Commiphora-Acacia* bushland, scattered *Terminalia*, *Euphorbia*, *Acacia*, etc. with shrub clumps, on both sandy and black soils; 15–620 m.

SYN. [*H. longiflorum* sensu Bak. & Wright in F.T.A. 4(2): 42 (1905), quoad *Hildebrandt* 2634 *non* (A.DC.) Jaub. & Spach.]

b. subsp. **harareense** (*S. Martins*) *Verdc.*, comb. et stat. nov. Type: Zimbabwe, Harare, Mabelreign, *Greatrex* in *G.H.* 26562 (K, holo.!, LISC, SRGH, iso.).

Annual or ephemeral unbranched or sparsely branched plant scarcely woody at base, 10–40 cm. tall. Leaves narrowly to broadly elliptic, 2.5–9.5 cm. long, 1–4 cm. wide, usually narrowly acuminate at apex, cuneate into very distinct petiole at the base. Calyx-lobes ± 2 mm. long. Corolla-tube ± 2.5 mm. long; lobes ± 1–1.5 × 1 mm. Fruits with asymmetrical sinus at apex of nutlets. Fig. 16/4, p. 60.

TANZANIA. Probably Mwanza, *R.L. Davis* 147!; Mpanda District: Rukwa valley, Milepa, Jan. 1949, *Burnett* 49/13! & 26 Dec. 1946, *Pielou* 49!; Mpwapwa, 5 Feb. 1930, *Hornby* 172!

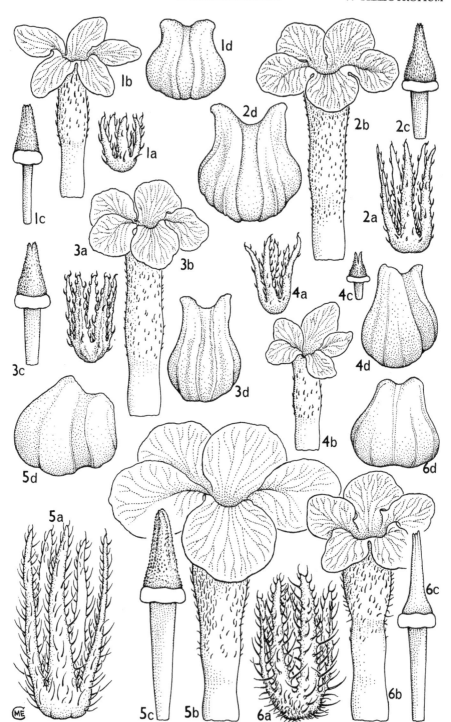

FIG. 16. *HELIOTROPIUM SPP.*—**a**, calyx, × 10; **b**, corolla, × 10; **c**, style and stigma, × 20; **d**, nutlet, × 12 of **1**, *HELIOTROPIUM PECTINATUM* subsp. *PECTINATUM*; **2**, *H. PECTINATUM* subsp. *SEPTENTRIONALE*; **3**, *H. PECTINATUM* subsp. *MKOMAZIENSE*; **4**, *H. PECTINATUM* subsp. *HARAREENSE*; **5**, *H. SCOTTEAE*; **6**, *H. SIMILE*. 1a–c, from *Faulkner* 1056; 1d, from *Stannard & Gilbert* 842; 2, from *Newbould* 6934; 3, from *Lye* 2404; 4a–c, from *Burnett* 49/13; 4d, from *Davis* 147; 5, from *Nattrass* 3508; 6a–c, from *Boppré* s.n.; 6d, from *Dummer* 4589. Drawn by Mrs Maureen Church.

DISTR. **T** ?1, 4, 5, 7 (see note); Malawi, Zimbabwe
HAB. Open grassy woodland; 700–995 m.

SYN. [*H. sudanicum* sensu F.W. Andrews in K.B. 8: 440 (1953), quoad *Hornby* 172 & *Davis* 147, *non* F.W.
    Andrews sensu stricto]
    H. hararense S. Martins in Garcia de Orta, Sér. Bot. 9: 74 (1988) & in F.Z. 7(4): 92, t. 26/9 (1990)

NOTE. *Bjørnstad* 1369 (Tanzania, Iringa District, Ruaha National Park, 3.5 km. NNW. of Msembe, 15
Feb. 1972) and *Backéus* 1096 (Iringa District, Mtera Resthouse, 9 Feb. 1975) have a slightly different
facies but belong here. Similar annual plants from Kenya, e.g. *R.B. & A.J. Faden* 74/1242 (Kenya,
Kilifi District, Sokoke Forest, 25 July 1974), clearly close to the type of *H. pectinatum* have longer
flowers but there is much variation. There is no doubt that subsp. *harareense* is very similar to *H.
sudanicum* F.W. Andrews in K.B. 8: 440 (1953), type: Sudan, Gezira Research Farm, *Ismail* J 45 (K,
holo.!, A, BR, FT, iso.).
    This was at first considered to be an introduced weed but I.M. Johnston who saw a duplicate of
the type reported he thought it was an indigenous African species related to *H. tiaridioides* Cham.
but differing from it in having a glabrous ovary and in lacking branched hairs. He was certain it
was not an introduced American plant. Both belong to the sect. *Coeloma*. I am not certain of the
relationship between true *H. sudanicum* and *H. pectinatum* but it might be best considered another
subspecies. I had at first intended to describe this as a subspecies under another name but there is
no doubt it is the same as Martin's taxon which behaves as a perfectly distinct species in Zimbabwe
where no related taxa occur.

    c. subsp. **septentrionale** *Verdc.*, a subsp. *pectinato* foliis majoribus usque 4 cm. latis, petiolis saepe
pro ratione longioribus. Typus: Northern Frontier Province, Ndoto, Ngurunit [Ngrunet], *Fratkin* 5 (K,
holo.!, EA, iso.!)

Annual ± unbranched herb or perennial with several stems 15–60 cm. tall and sometimes
spreading for 1 m. Leaves mostly broadly elliptic or ovate-lanceolate, 2–9 cm. long, 0.8–4 cm. wide, ±
rounded to acuminate at the apex, cuneate into a petiole up to 2–2.5 cm. long, up to ½ as long or even
as long as the lamina but often quite short. Calyx-lobes usually linear-lanceolate, 2–3 mm. long.
Corolla-tube 3–7 mm. long; lobes 1.5–3 mm. long, 1.5–2.5 mm. wide. Fruits with asymmetrical or very
symmetrical sinus at top of the nutlets which are sometimes distinctly 2-horned. Fig. 16/2.

KENYA. Northern Frontier Province: South Turkana, Lokori, 11 km. S. of Kangatet, 23 May 1970,
Mathew 6366! & 13 km. S. of Mado Gashi [Modo Gash] on Garissa road, 11 Dec. 1977, *Stannard &
Gilbert* 942!; Turkana District: Oropoi, *Newbould* 6934!
DISTR. **K** 1, 2; not known elsewhere
HAB. *Acacia tortilis* woodland, desert bushland and *Acacia, Euphorbia, Commiphora, Cordia*, etc.
bushland, dried river beds and also riverine vegetation; 290–900 m.

SYN. [*H. longiflorum* sensu Bak. & Wright in F.T.A. 4(2): 41 (1905), pro parte quoad *Wellby* s.n., *non*
    (A.DC.) Jaub. & Spach]

NOTE. *Curry & Glen* 41 (Northern Frontier Province, foot of Mt. Koiting, 31 Jan. 1969) in open
grassland at 1260 m. I at first kept separate; it has the calyx-lobes 3–5 mm. long, corolla-tube 6–7
mm. long with lobes 2.5–3 × 2.2–2.5 mm. and bristly long hairs on the stems, petioles and midrib;
leaves ovate-lanceolate, 7 × 2.2 cm., narrowly acute at the apex. Subsp. *septentrionale* is rather a
dustbin variant and this can well be included until further work indicates otherwise. The
inflorescence is somewhat like that of *H. scotteae* but the leaves are not at all what one would expect
of a hybrid.

    d. subsp. **mkomaziense** *Verdc.*, a subsp. *pectinato* foliis minoribus usque 4 × 1.2 cm. nuculis valde
bicornutis differt. Typus: Pare District, 13 km. NW. of Mkomazi, *Lye* 2404 (K, holo.!, MHU, iso.)

Perennial? herb to 40 cm., woody at the base. Leaves small, lanceolate, up to 4 × 1.2 cm., usually
smaller. Nutlets distinctly symmetrically 2-horned at the apex. Fig. 16/3.

TANZANIA. Pare District: 13 km. NW. of Mkomazi on Moshi–Korogwe road, 1 Apr. 1969, *Lye* 2404!
DISTR. **T** 3; known only from the type gathering
HAB. Dry scrubland; 600 m.

NOTE. The fruits of this are exactly those of *H. benadirense*. This variant is distinctive but there is little
to separate it from *H. pectinatum*. More material may, however, demonstrate it is constant in its
characters and deserves specific rank. No rootstock is preserved but it is woody and branched at
the base.

    7. **H. scotteae** *Rendle* in J.B. 70: 161 (1932); Kabuye & Agnew in U.K.W.F.: 520 (1974).
Types: Kenya, S. Nyeri District, Ragati R., near Karatina, *Rendle* 575, *Priestley & Scott* & N.
Nyeri District, Katherini R. and N. Rongai [Kasorongai] R. to W. Mount Kenya [Kenia]
Forest Station, *Mearns* 1254 (BM, syn.!)

Straggly or erect herb, sometimes almost climbing, 0.9–2.5 m. tall; stems sparsely to densely spreading hairy and with adpressed pubescence. Leaf-blades ovate to ovate-lanceolate, 3.5–15 cm. long, 2–10 cm. wide, acute at the apex, gradually narrowed at the base into the petiole, sometimes slightly bullate, ± velvety beneath when very young, later scabrid with ± sparse to dense tubercle-based hairs and many cystolith dots with or without minute hairs arising from them; petioles 1–2 cm. long. Inflorescences terminal, 2(–3)-branched, each scorpioid cyme 3–18 cm. long; peduncle 8–24 cm. long; flowers with strong ± unpleasant odour. Calyx divided ± to base, the lobes linear-subulate, up to 6 mm. long, setulose. Corolla-tube pale green or yellowish towards the base, 4–5 mm. long, slightly contracted about the middle, the upper ⅔ pubescent outside; throat yellow; limb white or greenish cream with a pale green vein extending halfway to outer edge of the lobe, yellow at junction of tube and limb; lobes ± round, obovate-oblong or subreniform, 2.5–3 mm. long, 1–2.5 mm. wide, narrowed at the base, glabrous. Stamens with anthers ± 2 mm. long affixed at about the middle of the tube; filaments very short. Style green, 1.25 mm. long with a stigmatose ring at junction with stigmatic cone, the latter 1.25 mm. long, papillate and shortly 4-lobed at the apex. Nutlets brown, subovoid, 2.5 mm. long and wide, with a distinct asymmetrical groove and an indistinct one, glabrous. Fig. 16/5, p. 60; 17.

KENYA. Northern Frontier Province: Mathews Range [Ol Doinyo Lengiyo], 20 Dec. 1958, *Newbould* 3300!; Embu, 11 Apr. 1932, *Sunman* in *A.D.* 2224!; Machakos District: Ol Doinyo Sapuk, 26 Jan. 1964, *Napper et al.* 1715!
DISTR. **K** 1, ?3, 4; not known elsewhere
HAB. *Juniperus* forest, forest clearings and bushland; 1500–2100 m.

8. **H. simile** *Vatke* in Linnaea 43: 317 (1882); Gürke in P.O.A. C: 337 (1895); Bak. & Wright in F.T.A. 4(2): 33 (1905). Type: Kenya, Ukamba, *Hildebrandt* 2849 (B, holo.†)

Prostrate or ascending annual or perennial herb 7–90 cm. tall; stems, petioles and inflorescences pubescent with long bristly and shorter finer ± spreading mixed white hairs, often quite dense. Leaves alternate or ± opposite, blades mostly ovate, elliptic or oblong, 1–11.5 cm. long, 1–5.5(–7.5) cm. wide, acute to rounded at the apex, cuneate, rounded or ± truncate at the base, obscurely to fairly distinctly crenate or undulate, pubescent with bulbous-based and finer hairs above, ± densely ± adpressed pubescent to sometimes quite woolly beneath, the hairs on the venation longest and the finer hairs with their tubercles giving a granular appearance to the surface, sometimes slightly bullate with the varying direction of the indumentum adding emphasis; petioles 0.8–4.5 cm. long. Inflorescences unbranched or 2–3-branched, the young cymes 2 cm. long extending to 23 cm. with numerous very closely placed flowers; peduncles 4–16 cm. long. Calyx-lobes linear to lanceolate or oblong-lanceolate from a narrowly triangular base, often a little unequal, 2–5 mm. long, spreading pubescent. Corolla white; tube 3–6 mm. long, mostly inflated about the middle, ± densely pubescent outside; lobes often quite unequal, obovate-oblong, 1.5–4 mm. long, 1.2–2 mm. wide, rounded at apex. Style 1.5 mm. long; stigmatic cone 1.5 mm. long, narrow and papillate. Fruit depressed globose, 2 mm. tall, 3–4 mm. wide, with median groove narrow and not extending to base of nutlets but with a commissure where they join at apex so that nutlets are undulate but not 2-horned at apex; each nutlet 3-locular in section. Fig. 16/6, p. 60.

KENYA. Northern Frontier Province: S. Turkana, 16 km. from Lokori on road to Sigor, 28 May 1970, *Mathew* 6430!; Turkana District: 20 km. NW. of Lomoru Itae [Itale] on road to Kaiemothia, 3 Nov. 1977, *Carter & Stannard* 163b!; Tana River District: Kora Game Reserve, 13 May 1983, *Mungai et al.* 112!
TANZANIA. Moshi District: Himo, 21 Oct. 1981, *Archbold* 2958!; Lushoto District: W. Usambara Mts., N. of Mombo, 8 July 1960, *Leach & Brunton* 10207! & Umba valley, Dec. 1896, *C.S. Smith*! & above Mombo, 8 July 1932, *Geilinger* 834!
DISTR. **K** 1, 2, 4, 7; **T** 2, 3; ?Ethiopia
HAB. Bushland and scrub, particularly *Euphorbia*, *Acacia*, *Commiphora* and *Albizia* especially on lava-flows, but also on orange-red loams and gritty sandy soil over basement rocks; also roadsides and occasionally cultivations; (150–)420–1000 m.

SYN. [*H. indicum* sensu Bak. & Wright in F.T.A. 4(2): 33 (1905) pro parte quoad *C.S. Smith* s.n. *non* L.]

NOTE. Grows together with and has been confused with *H. steudneri* Vatke (e.g. Shimba Hills, near Kwale, June 1982, *Boppré* s.n.) but easily distinguishable by the longer petioles.

9. **H. steudneri** *Vatke* in Oest. Bot. Zeitschr. 25: 167 (1875) & in Linnaea 43: 318, 320 (1882); Engl., Hochgebirgsfl. Trop. Afr.: 352 (1892); Schweinf. in Ghiká, Pays des Somalis:

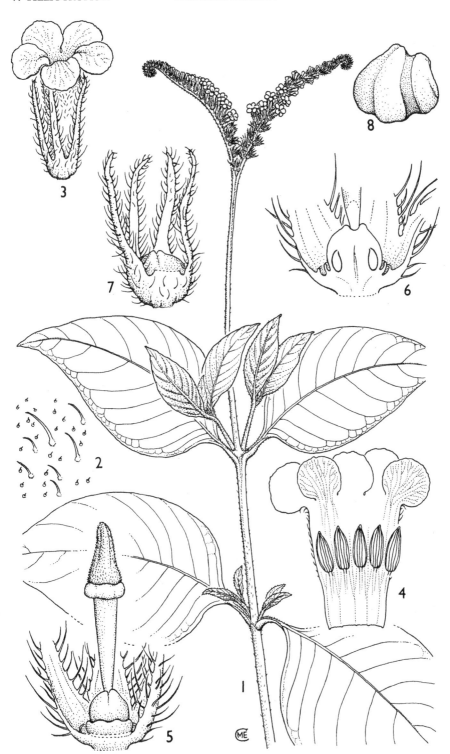

FIG. 17. *HELIOTROPIUM SCOTTEAE*—**1**, habit, × ⅔; **2**, part of upper leaf surface, × 12; **3**, flower, × 5; **4**, corolla, opened out, × 6; **5**, gynoecium, × 16; **6**, ovary, longitudinal section, × 20; **7**, young fruit, × 6; **8**, nutlet, × 10. 1, 2, from *Bopprè* 7; 3–8, from *Nattrass* 3508. Drawn by Mrs Maureen Church.

211 (1898); Bak. & Rendle in F.T.A. 4(2): 42 (1905); F.P.S. 3: 87 (1956); E.P.A.: 776 (1962); Kabuye & Agnew in U.K.W.F.: 520 (1974). Types: Ethiopia, Eritrea, Bogos, *Hildebrandt* 469 (B, syn.†, BM, isosyn.!) & Keren, *Beccari* 145 (B, syn.†, K, isosyn.!)

Perennial ± erect or spreading herb, sometimes subshrubby, 0.15–1 m. tall; stems branched, usually distinctly woody below, several from a woody rootstock, densely ± persistently hairy with spreading and short ± adpressed white hairs, or all adpressed; lower stems with ± glabrous fissured epidermis. Leaf-blades narrowly elliptic to elliptic-lanceolate or oblong, 0.7–9(–12.5) cm. long, 0.4–2.5(–3.2) cm. wide, ± acute at the apex, cuneate at the base, often obscurely crenulate and sometimes distinctly wrinkled, frequently distinctly bullate above with venation impressed, sparsely to very densely covered with long and short white hairs above and mainly on raised nerves beneath with rest of lower surface with only much shorter hairs or ± granular; petiole mostly short (to ± 1.2 cm.), only about $\frac{1}{10}$ the length of the lamina or even obsolete. Cymes mostly long and simple but often 2–3-branched, very short and scorpioid at first, soon extending to 5–20(–37) cm. long, the peduncle 1.5–7 cm. long, the axes with long and short hairs; flowers sessile, often foul-smelling. Calyx very variable, 2.5–4.5(–5.5) mm. long, densely pubescent, lobed ± to the base, the lobes linear-lanceolate, 0.5 mm. wide, tapering. Corolla white or creamy yellow with yellow to deep olive centre; tube narrowly funnel-shaped above and pubescent, narrower and glabrous at the base, 4–5.5 mm. long; lobes obovate-oblong or oblong, 2–3 mm. long, 1.5–1.8 mm. wide, rounded at apex, strongly reticulate veined. Anthers variably placed in upper middle part of the tube or even below the middle but tips not exserted. Style (0.5–)0.6–2 mm. long; stigma conical with narrowed tip, 0.8–1.8 mm. long, papillate with short stiff hairs, often shortly bilobed at the tip, each lobe again shortly bifid. Fruit depressed ovoid, 2.5–3 mm. tall, 3.5 mm. wide, glabrous, slightly rugulose. Nutlets 2, somewhat flanged where they join and with an asymmetrical shallow narrow groove from apex to base and often lateral elliptic depressions on the junction.

KEY TO INFRASPECIFIC VARIANTS

1. Indumentum adpressed and not at all woolly; leaves less or
     not bullate (subsp. **steudneri**)   2
   Indumentum spreading and woolly; leaves ± bullate
     (Uganda material has shorter indumentum)   .   .   .   subsp. **bullatum**
2. Inflorescences 5–20 cm. long   .   .   .   .   var. **steudneri**
   Inflorescences up to over 35 cm. long (**T** 7)   .   .   .   .   var. **iringensis**

subsp. **steudneri**

Indumentum adpressed and not at all woolly. Leaves mostly less bullate or quite plane.

var. **steudneri**

Leaves scarcely if at all bullate. Inflorescences 5–20 cm. long, usually quite short.

KENYA. Machakos District: Tsavo Park West, just S. of Ngulia Hills, 4 Aug. 1963, *Verdcourt* 3695! & Masalani, 23 Apr. 1969, *Napper & Kanuri* 2054!; Teita District: 12 km. SW. of Voi on Taveta road, 11 Feb. 1966, *Gillett & Burtt* 17172!
TANZANIA. Pare District: Same, 1 Sept. 1969, *Batty* 591!; Lushoto District, 3 km. WSW. of Mkomazi on Moshi road, 28 Mar. 1975, *Hooper & Townsend* 1033! & Mkumbara, 13 Oct. 1971, *Magogo* 137!
DISTR. **K** 1, 4, 7; **T** 3, 5–7 (see note); Ethiopia
HAB. Bushland, particularly *Combretum-Acacia-Commiphora* and *Boscia-Strychnos-Delonix-Combretum*, etc., also grassland; (± ?0–)200–900 m.

SYN. *H. eduardii* Martelli, Fl. Bogos.: 59 (1886); Bak. & Wright in F.T.A. 4(2): 43 (1905). Type: Ethiopia, Eritrea, Keren, *Beccari* 145 (FT, holo., K, iso.!)
     [*H. indicum* sensu Bak. & Wright in F.T.A. 4(2): 32 (1905), pro parte, quoad specim. *Wakefield*, non L.]
     [*H. longiflorum* sensu Bak. & Wright in F.T.A. 4(2): 41 (1905), pro parte, quoad *Goetze* 416, *non (A.DC.)* Jaub. & Spach]

---

* This had been correctly named *H. steudneri* at B, probably by comparison with holotype.

NOTE. A number of specimens from low altitudes, even near sea-level, seem better placed in *H. steudneri* than *H. pectinatum* by virtue of short petioles, ± crenulate leaves and indumentum, e.g. *Verdcourt* 1190 (Kenya, Malindi, 26 Dec. 1954), *Rawlins* 129 (Kenya, Lamu District, Kui I., Sept. 1956) (I know that mine at least was correctly localised). *Rawlins* s.n. from the same locality, June 1956 is a curious plant with opposite leaves. A number of other specimens are better placed in *H. steudneri*, e.g. *Bally* 8649 (Kenya, Voi, Mazinga Hill, 1 Feb. 1953), but the petioles are longer than usual. The basis of the **T** 5, 6 and 7 records *Cribb et al.* 11224 (Iringa District, Mikumi to Iringa, 30 Jan. 1979) *Anderson* 571 (Mpwapwa District, Kongwa, 10 Jan. 1950) and *Goetze* 416 (Kilosa District, Vidunda, 1 Jan. 1899) have longer petioles and are not typical. *Perdue & Kibuwa* 11001 (Kilosa District, 65.6 km. W. of Mikami (? Mikumi), 8 Aug. 1971) is rather similar to *Anderson* 571 but has leaves up to 14.5 × 4 cm.; possibly a distinct variant.

var. **iringensis** *Verdc.*, var. nov. a var. *steudneri* inflorescentiis valde elongatis usque 37 cm. longis, tubo corollae ± 5 mm. longo differt. Typus: Tanzania, Iringa District, Kidatu, *Mhoro* 167b (K, holo.!)

Branched shrub with ± bullate leaves and very raised venation beneath. Inflorescences longer than in other variants, up to 37 cm. long. Corolla-tube ± 5 mm. long, finely pubescent.

TANZANIA. Iringa District: Kidatu, 14 Jan. 1971, *Mhoro* 167b!
DISTR. **T** 7; known only from the type gathering
HAB. Riverside on sandy soil; between 200 and 500 m.

NOTE. Differs from *H. pectinatum* in the fine adpressed indumentum on the stems and leaves and lack of stiff bristly hairs save for short ones on calyx. It was first considered this might be a distinct species but despite the very long inflorescences there is nothing in the actual floral structure etc. More material is required. *Ludanga* in *M.R.C.* 3271 (**T** 6, Ruaha R.) and *Cribb et al.* 11224 (**T** 7, Mikumi–Iringa, by Ruaha R.) are intermediate.

subsp. **bullatum** *Verdc.*, subsp. nov. a subsp. *steudneri* foliis bullatis dense patenter lanatis indumento caulium patenti differt. Typus: Kenya, Masai District, Ngong Hills, near Kekonyokie, *Verdcourt* 3668 (K, holo.!, EA, iso.!)

Indumentum spreading and ± woolly. Leaves ± bullate. Fig. 18/1, p. 66.

UGANDA. Karamoja District: Moroto, 2 Jan. 1937, *A.S. Thomas* 2134! & 21 May 1940, *A.S. Thomas* 3420! & Moroto, Kasineri Estate, July 1971, *J. Wilson* 2032!
KENYA. Naivasha District: near Longonot, 29 May 1950, *Le Pelley* 3500!; Nairobi District: Thika Road House, 16 May 1951, *Verdcourt* 511!; Masai District: Namanga –Amboseli, about 16 km. from Ol Tukai, 14 Dec. 1959, *Verdcourt* 2564!
TANZANIA. Musoma District: km. 8 on Seronera to Magadi Lake road, 12 May 1961, *Greenway* 10174!; Arusha District: about 3 km. S. of Ol Doinyo Sambu, 1 Nov. 1955, *Milne-Redhead & Taylor* 7023!; Pare District: on Pangani R., between Arusha Chini and Same, 57 km. from Moshi, Nyumba ya Mungu, 14 Aug. 1968, *Batty* 269!
DISTR. **U** 1; **K** 1–4, 6; **T** 1–3, 7; not known elsewhere
HAB. Grassland, *Acacia-Commiphora* woodland and bushland, roadsides, sometimes on black cotton soil, also occasionally as a weed in newly derived cultivations; 750–2070 m.

SYN. [*H. steudneri* sensu Kabuye & Agnew in U.K.W.F.: 520, fig. on p. 519 (1974), pro parte, *non* Vatke sensu stricto]

NOTE. A few specimens, e.g. *Lamprey* 262 (Mbulu District, Tarangire R., 7 Dec. 1957 at 1140 m.) approach subsp. *steudneri*. The Uganda material has shorter indumentum but has been retained here. *T. & B. Poćs* 87235! (Tanzania, Masai District, Lake Natron Basin, Engare Sero Gorge near Garaselo, 14 Dec. 1987) has rounded leaf-apices and adpressed velvety stems but I think belongs here rather than with var. *steudneri*.

NOTE. (on species as a whole) Bak. & Wright keep *H. eduardii* and *H. steudneri* separate but cite *Beccari* 145 under both! B.L. Burtt (adnot.) compared the types and thought them conspecific and I agree. Ovoid densely woolly flower galls are common but the cause is not known. The S. African *H. nelsonii* C.H. Wright in Fl. Cap. 4(2): 9 (1904) (South Africa, Transvaal, bend of Vaal R., *Nelson* 219 (K, lecto.!)), and its synonym, *H. dissimile* N.E. Br., has often been sunk into *H. steudneri* (e.g. by Martins in F.Z. 7(4): 92, t. 26/10 (1990)) and probably represents a third subspecies; it certainly is not identical with the Eritrean type. In a broad sense, including *H. nelsonii*, *H. steudneri* extends to Mozambique, Zimbabwe, Botswana, SW. Angola, Namibia and South Africa. The evil smell is linked to pollination by flies.

10. **H. sp. A**

Perennial herb 15–25 cm. tall, branched from or just above the base; stems woody at base, dark brown, ± ridged, ± glabrous, paler above and with dense curled ± buff hairs, particularly on youngest parts. Leaves elliptic, 1.5–5 cm. long, 0.6–2 cm. wide, rounded or subacute at the apex, cuneate at the base into a petiole up to 8 mm. long, almost glabrous to sparsely pubescent with short adpressed hairs, particularly on the sunken midrib above

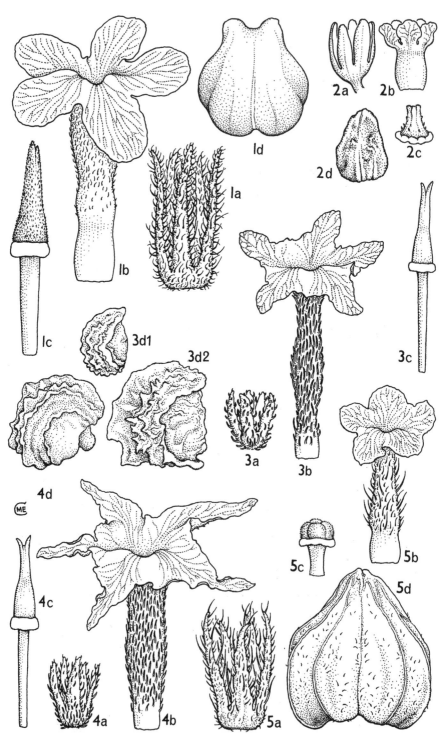

FIG. 18. *HELIOTROPIUM SPP.*—**a**, calyx, × 10; **b**, corolla, × 10; **c**, style and stigma, × 20; **d**, nutlet, × 12 of **1**, *H. STEUDNERI* subsp. *BULLATUM*; **2**, *H. CURASSAVICUM*; **3**, *H. LONGIFLORUM* var. *STENOPHYLLUM*; **4**, *H. LONGIFLORUM* subsp. *UNDULATIFOLIUM*; **5**, *H. INDICUM*. 1, from *Greenway & Kanuri* 12393, 2, from *Richards* 21593; 3a–d1, from *Martin* 139; 3d2, from *Haylett* 11; 4, from *Bally* 12278; 5, from *Greenway & Kanuri* 14732. Drawn by Mrs Maureen Church.

and with curled hairs on margin and ± raised nervation beneath but rest of undersurface merely granular. Inflorescences condensed, up to 5-branched, the branches cincinnate, short, up to 3 cm. long (probably longer at maturity); axes densely pubescent like the young stems. Flowers crowded, 5–6-merous. Calyx pubescent; lobes rather unequal, linear, 3 mm. long, 0.25–0.5 mm. wide, with dense short white ascending hairs. Corolla bright white; tube 4 mm. long, swollen just below the throat and slightly at extreme base, pubescent save at base; lobes oblong, 1.5 mm. long, 0.8–1.2 mm. wide, crinkly. Anthers 1.8 mm. long, narrowed to acuminate apex. Stigma pale, narrowly conical, 1.5 mm. long, minutely bifid, papillate-puberulous; style dark, 1.3 mm. long. Fruit not seen.

KENYA. Tana River District: 4 km. N. of SKT 15, 1°11′S, 39°51′E, 21 Dec. 1964, *Gillett* 16522!
DISTR. **K** 7; not known elsewhere
HAB. Grassland with scattered *Acacia* on clay soil; 96 m.

11. **H. curassavicum** *L.*, Sp. Pl.: 130 (1753); I.M. Johnston in Contr. Gray Herb. 81: 14 (1928) & in Lundell et al., Fl. Texas 1: 145 (1964, rpt. 1967); Correll & M.C. Johnston, Man. Vasc. Pl. Texas: 1288 (1970); Kazmi in Journ. Arn. Arb. 51: 179 (1970); Brummitt in Fl. Europaea 3: 86 (1972); Riedl in Fl. Turkey 6: 255, fig. 11/3 (1978); Martins in F.Z. 7(4): 93, t. 26/12 (1990). Type: Curaçao, Morris., Pl. Hist. etc. 3: 452, S. 11, t. 31/12 (1699)*

Branched prostrate glabrous somewhat fleshy ± glaucous perennial herb with radiating stems 7–20(–40) cm. long, usually from a deep rhizome or tuber; flowering stems ± ascending. Leaves oblanceolate or oblong, 1–5 cm. long, 3–8(–20) mm. wide, obtuse or rounded at the apex, narrowed at the base, ± succulent and juicy. Inflorescences terminal or extra-axillary and lateral along the leafy stems, single, paired or rarely ternate, densely-flowered scorpioid cymes, elongating in fruit to up to 10 cm.; bracts absent; pedicels obsolete. Calyx lobed almost to the base; lobes ± equal, lanceolate or oblong, 1–1.4 mm. long, ± obtuse, ± accrescent in fruit, fleshy. Corolla white, mostly with a greenish yellow eye and sometimes a faint pink stripe on the lobes; tube cylindrical, 1–1.5 mm. long, 0.8–1.2 mm. diameter, slightly shorter than the calyx-lobes; limb 1.5–3(–10) mm. diameter, the lobes rounded, 0.6–1 mm. long, 0.8–1.3 mm. wide, ascending or somewhat spreading. Anthers lanceolate, pointed, not cohering, held about 0.5 mm. above the corolla base, 0.8–1 mm. long, subsessile. Stigma 0.5 mm. long, 0.6 mm. wide, conic with a truncate obscurely 4-lobulate apex, becoming discoid and somewhat 4–5-sided in fruit. Fruit 4-lobed, 2–2.5 mm. tall, 2–3 mm. wide, smooth; nutlets 4, obscurely didymous, the backs smooth and rounded, ventrally angled, rough; each nutlet 1-seeded with a bony endocarp and dorsally with a thick layer of firm vesicular exocarp which acts as a float for water dissemination.

var. **curassavicum**

Corolla-limb 1.5–3 mm. wide; leaves up to 8 mm. wide. Fig. 18/2.

TANZANIA. Lushoto District: Mkomazi, Lake Manka, 26 July 1969, *Archbold* 1081! & same locality, 12 Jan. 1967, *Richards* 21953! & Mkomazi R., by Korogwe–Same road, 5 Apr. 1972, *Wingfield* 1927!
DISTR. **T** 3; Mozambique, Botswana and Angola; native of U.S.A. (Florida, Oklahoma, Texas and New Mexico), W. Indies, Mexico and Central America to Surinam and Colombia thence along Pacific Coast to Chile and Patagonia; introduced into several areas of the Old World, S. Europe, Turkey, India, Ceylon, Pacific Is., etc.; widely cultivated in Europe and Asia and becoming naturalised
HAB. Gregarious at muddy stream edges and alkaline lake edges; 435–480 m.

SYN. *Coldenia succulenta* A. Peter, in Abh. Ges. Wiss. Göttingen, N.F., 13(2): 90 (1928). Type: Tanzania, Mkomazi [Mkomasi], Lake Manka [Manga-See], *Peter* 10857 (B, holo.!)**

NOTE. The much larger flowered var. *obovatum* DC. (*H. spathulatum* Rydb.) is restricted to the New World and has the corolla 5–10 mm. wide.

12. **H. longiflorum** (*A.DC.*) *Jaub. & Spach*, Ill. Pl. Or. 4: 96, t. 360 (1852); Bunge in Bull. Soc. Imp. Nat. Moscou 42: 304 (1869); Bak. & Wright, F.T.A. 4(2): 41 (1905); Blatter in Rec. Bot. Surv. India 8: 310 (1921); Schwartz in Mitt. Inst. Bot. Hamb. 10: 209 (1939); F.P.S. 3: 87

* I have accepted I.M. Johnston's selection of type locality as indicating he chose this as lectotype from the several syntypes cited by Linnaeus
** The identity of this was established in 1974 by Richardson, University of Texas long before I independently guessed its identity from Peters' description and type locality (see Rhodora 79: 568 (1977)).

(1956); E.P.A.: 774 (1962). Type: Arabia, Djebel Sidr [Mt. Sedder], *Schimper* 842 (G, holo., K, isotypes!)

Annual or perennial herb or subshrub 5–60 cm. tall from a thin to thick tap-root or clump-forming subshrub to 60 cm. wide from a very woody base; stems little–much-branched, densely covered with absolutely adpressed white hairs. Leaves aromatic, with goat-like smell when dry; blades linear-lanceolate, linear-oblong or elliptic-oblong or distinctly ovate to broadly elliptic, 1–8 cm. long, 0.1–3.5 cm. wide, subobtuse to acute at the apex, ± rounded to cuneate at the base, adpressed white setulose pubescent on both surfaces but mainly on nerves beneath and densely dotted with non-hair-bearing cystoliths, the margins often distinctly undulate; nerves impressed above; petiole ± 0–2 cm. long. Cymes distinctly scorpioid, mostly short at first, 1.5–2.5 cm. long, lengthening to 10–15 cm. in fruit; peduncle ± 3–4 cm. long; flowers sessile. Calyx-lobes linear to lanceolate or oblong, 1–1.5 mm. long, 0.2 mm. wide. Corolla pure white or cream or sometimes with yellowish centre; tube cylindrical but usually distinctly swollen in middle and fusiform in dry state, 3.5–6 mm. long, usually densely adpressed pubescent, but sometimes ± glabrous; limb 4–7 mm. wide, the lobes short and blunt to long and narrowly triangular, 0.5–3 mm. long, 1–2.5 mm. wide. Style and stigma together 1.5–3 mm. long, almost equal in length or style longer; stigmatic club narrowly conical, widened at the base, shortly 2–4-fid at the apex. Fruit 2–3 mm. long, 3 mm. wide, divided into 2–4 nutlets; nutlets segment-shaped, the margins variable, sometimes slightly to very distinctly winged, the wings often crenate, or merely rugose or nodulose or even serrate, outer curved face between the wings rugose to reticulate-echinate, the flat face with winged nodular or rugose ridge just below the marginal wing and area beyond smooth.

subsp. **longiflorum**

Annual or perennial herb or subshrub. Leaves linear to distinctly ovate or broadly elliptic. Inflorescences often longer. Corolla-limb with short obtuse lobes.

SYN. *Heliophytum longiflorum* A.DC. in DC., Prodr. 9: 555 (1845)

var. **stenophyllum** *Schwartz* in Mitt. Inst. Bot. Hamb. 10: 209 (1939). Type*: Yemen, Menâcha, *Schweinfurth* 1557 (?HBG, syn., K, isosyn.!)

Leaves linear to linear-lanceolate, mostly 1–10 mm. wide, rarely up to 1.5 cm. wide in a few specimens. Fig.18/3, p. 66.

UGANDA. Karamoja District: Eastern Matheniko, Oct. 1958, *J. Wilson* 627!
KENYA. Northern Frontier Province: Samburu Game Reserve, about 6 km. W. of Samburu Lodge, 5 Apr. 1977, *Hooper & Townsend* 1663! & South Horr, 10 km. S. where valley begins, 12 Nov. 1978, *Hepper & Jaeger* 6758!; Turkana District: 80 km. W. of Lodwar, 11 May 1953, *Padwa* 131!
DISTR. U 1; K 1–3; Arabia, Sudan, Ethiopia, Somalia
HAB. Desert grassland on sandy or lava plains, grassland with *Hyphaene*, etc., open bushland and *Acacia* woodland; 300–1200 m.

SYN. *H. somalense* Vatke in Oest. Bot. Zeitschr. 25: 166 (1875); Bak. & Wright in F.T.A. 4(2): 43 (1905); E.P.A.: 776 (1962). Type: Somalia (N.), Ahl Mts., Damalle, *Hildebrandt* 846 B (B, holo.†)

NOTE. Var. *longiflorum* with much wider leaves seems to occur only in Arabia but there the two varieties merge. *H. graminifolium* Chiov. (Fl. Somala 2: 319 (1932); E.P.A.: 774 (1962); type: Somalia (S.), Mogadishu, *Senni* 592 (FT, holo.!)), with sessile leaves and distinctive habit, is probably only a dune variant but further material is needed.

subsp. **undulatifolium** (*Turrill*) *Verdc.*, comb. & stat. nov. Type: Kenya, Masai District, open plains beyond Uaso Nyiro [Guaso Nyoro], *M.S. Evans* 767 (K, lecto.!)

Subshrubby with woody rootstock, often forming clumps. Leaves linear to linear-lanceolate, always under 1.2 cm. wide, often with very distinctly undulate margins. Inflorescences mostly short. Corolla-limb with distinctly longer narrowly triangular acute lobes up to 3 mm. long. Fig. 18/4, p. 66.

KENYA. Naivasha, 29 May 1950, *Le Pelley* 3501!; Laikipia, Ngobit, 22 Apr. 1952, *Bally* 8192!; Masai District: 8 km. on Narok–Nairobi road, 11 Dec. 1963, *Verdcourt* 3821!
TANZANIA. Shinyanga, road between Sakamaliwa [Sakamalewa] and Sekenke, Uduhe, 24 Jan. 1936, *B.D. Burtt* 5524!; Masai District: Serengeti, Olongogo to Olbalbal, 30 Jan. 1962, *Newbould* 5916!; Singida District: Iramba Plateau, Kisiriri [Kasiriri], 29 Apr. 1962, *Polhill & Paulo* 2241!
DISTR. K 1 (Isiolo area), 3, 4 (Nairobi), 5 (rare) 6, 7 (Tsavo); T 1–3, 5; not known elsewhere

---

* Schwartz cites 9 syntypes; the other 8 I have not seen.

Hab. Open grassland on sandy and volcanic plains, also open places in *Tarchonanthus* scrub, scattered *Acacia* grassland, *Acacia-Hyphaene-Sporobolus* associations etc. and remains as a weed in newly cultivated areas and persists in old cultivations; particularly common by roadsides; (30–) 450–2100(–2670) m., see note

Syn. *H. engleri* F. Vaupel in E.J. 48: 535 (1912). Types: Kenya, Lake Nakuru, *Engler* 1988 & 2071 (B, syn.†)
　　　H. undulatifolium Turrill in K.B. 1915: 76 (1915); Ivens, E. Afr. Weeds: 81 (1967); Kabuye & Agnew in U.K.W.F.: 520 (1974)

Note. This common upland plant sometimes occurs in white sheets and has an unpleasant smell reminiscent of goats. The highland population from about 1800–2100 m. is very uniform but some intermediates with subsp. *longiflorum* do occur. *Stannard & Gilbert* 873 (Kenya, Northern Frontier Province, Isiolo–Wajir road 16 km. from Mado Goshi [Modo Gash]) has the corolla-lobes of subsp. *undulatifolium* but comes from an area where the typical subspecies occurs. A puzzling specimen is *R.B. & A.J. Faden* 74/1056 (Kenya, Tana River District, Galole–Garsen, 8 km. along road towards Garsen from turn-off to Wenje) which has the distinct pointed corolla-lobes but is from an altitude of only 30 m.; it closely resembles *H. graminifolium* Chiov. in habit but that has short obtuse lobes. The variation in fruit cuts across the nomenclature adopted here. I think all variants can sporadically have winged or non-winged nutlets — in fact in some cases they may be growth stages. The winged form of subsp. *longiflora* is var. *lophocarpa* Jaub. & Spach, Ill. Pl. Or. 4: 97, t. 361 (1852); type: Saudi Arabia, Taifa, *Botta* (P, holo.). The S. African *H. lineare* (A.DC.) C.H. Wright (*H. kuntzei* Gürke) is closely related. It must be noted that if subsp. *undulatifolium* is maintained as a species it must be called *H. engleri*.

13. **H. indicum** L., Sp. Pl.: 130 (1753); Sims in Bot. Mag. 43, t. 1837 (1816) (early refs.); Vatke in Linnaea 43: 318 (1882); C.B.Cl. in Fl. Brit. Ind. 4: 152 (1883); Hiern, Cat. Afr. Pl. Welw. 1: 719 (1898); Bak. & Wright in F.T.A. 4(2): 32 (1905); I.M. Johnston in Contr. Gray Herb. 81: 19 (1928) & in Journ. Arn. Arb. 32: 111 (1951); F.P.S. 3: 85 (1956); Heine in F.W.T.A., ed. 2, 2: 321 (1963); Taton, F.C.B., Borag.: 28, map 8 (1971); Vollesen in Opera Bot. 59: 76 (1980); Martins in F.Z. 7(4): 91, t. 26/8 (1990). Type: Ceylon, *Hermann* 4.67, 1.9 (BM-HERM, syn.!)

Robust erect tufted perennial or annual herb or subshrubby herb 0.2–1.5 m. tall, usually much branched; stems woody at the base; young stems juicy, with short pubescence and coarse bristly longer white hairs, eventually glabrous or obscurely tuberculate. Leaves alternate or opposite, blades ovate to elliptic or ± triangular, (1.5–)3–16 cm. long, (0.5–)1.3–10 cm. wide, acute or acuminate at the apex, mostly abruptly truncate then very narrowly attenuate into the 1–7 cm. long petiole, or even ± subcordate or obtuse, the margin irregularly undulate, slightly bullate with nervation distinct beneath, with dense short tubercle-based hairs and sparser longer white hairs, sometimes almost velvety beneath. Flowers numerous in long dense single scorpioid cymes, unilateral, 2.5–45 cm. long, rolled up at the summit, with short pubescence and long setiform white hairs; peduncle up to 5 cm. long; pedicels obsolete. Calyx divided ± to the base into lanceolate leafy lobes 2.5–3 mm. long, sparsely white bristly hairy and with shorter pubescence. Corolla blue, lilac or often white; tube cylindrical, 3–4.5 mm. long, widened at the base, funnel-shaped above, sparsely pubescent outside; limb 3–4 mm. wide, lobed for ± 1 mm., the lobes rounded, 0.5–1 mm. long, 1–1.5 mm. wide. Ovary globose with 4 erect fleshy crests; style 0.5–0.7 mm. long, the stigma discoid surmounted by a hemispherical dome. Fruits brown, ovoid, glabrous or glabrescent, grooved, splitting first into 2 cohering pairs of nutlets then into 4 trigonous-ovoid nutlets 4 mm. long, 2 mm. wide, produced, almost wing-like, and acuminate at the apex, the outer face with 3 sharp ridges ± anastomosing at the apex; internal faces devoid of cavities, smooth and convex. Fig. 18/5, p. 66.

Tanzania. Tanga District: Muheza [Muhesa], 17 Nov. 1929, *Greenway* 1884!; Rufiji District: Rufiji R., S. bank, Utete, 2 Dec. 1955, *Milne-Redhead & Taylor* 7461!; Iringa District: Great Ruaha River ferry, 8 Dec. 1970, *Greenway & Kanuri* 14732!
Distr. T 3–8; Senegal to Angola, Zaire, Burundi, Ethiopia, Sudan, Mozambique, Malawi, Zambia, Zimbabwe, Caprivi Strip, South Africa, Namibia, N. Argentina to S. U.S.A. and throughout the tropics of Old World, probably originally an American species
Hab. Bushland, plantations, old cultivations, roadsides, mostly in damp sandy places, e.g. river banks, also on riverine mud; 0–975(–1600) m.

Syn. *H. africanum* Schumach. & Thonn. in Schumach., Beskr. Guin. Pl: 87 (1827); DC., Prodr. 9: 548 (1845); Bak. & Wright in F.T.A. 4(2): 43 (1905); F.W.T.A. 2: 199 (1931). Type: Ghana [Guinea], *Thonning* 69 (C, holo.)
　　　Heliophytum indicum (L.) DC., Prodr. 9: 556 (1845)

NOTE. A small ephemeral 9 cm. tall (Tanzania, shore of Great Ruaha R., 19 Oct. 1969 at 900 m., *Batty* 688!) is I am sure a starved form of this flowering while almost a seedling.

All the East African material seen appears to be true *H. indicum* L. but another very closely allied S. American plant *H. elongatum* Hoffm. (*H. decipiens* Backer) has been recorded from Indonesia and Zimbabwe. Van Steenis claimed one could tell the two from a moving train but they are certainly very closely similar in facies. In *H. indicum* the usually smaller nutlets diverge considerably at maturity and are distinctly bifid at the apex whereas in *H. elongatum* the divergence is less marked and they are only slightly notched. Full details can be found in I.M. Johnston in Contr. Gray Herb. 81: 18 (1928) and van Steenis in Bull. Inst. Bot. Buitenz., sér. 3, 13: 286–7 (1934) (the fruit figures in the latter are so poor I suspect both were drawn from specimens of *H. indicum*). The illustrations 514 and 515 in Backer, Atlas of 220 Weeds of Sugarcane Fields in Java (1973) are helpful.

14. **H. rariflorum** *Stocks* in Hook., Journ. Bot. 4: 174 (1852); Boiss., Fl. Orient. 4: 144 (1875); C.B. Cl. in Fl. Brit. Ind. 4: 152 (1883); Bak. & Wright in F.T.A. 4(2): 40 (1905); Cook, Fl. Bombay 2: 209 (1908); Burkill, Fl. Pl. Baluchistan : 50 (1909); F.P.S. 3: 88 (1956); E.P.A.: 776 (1962); Heine in K.B. 16: 203 (1962); Riedl in Fl. Iran. 48/15: 15 (1967); Kazmi in Journ. Arn. Arb. 51: 151 (1970); Kabuye & Agnew in U.K.W.F.: 518 (1974). Type: Pakistan, Karachi [Kurrachi], Jemadar Ka Landa, *Stocks* 492 (K, holo.!)

Much-branched woody subshrub or subshrubby herb 20–80 cm. tall; rootstock ± simple, often very woody, up to 1 cm. diameter, but usually more slender; young branches densely white strigose, older with silvery epidermis peeling to reveal dark brown bark which eventually also peels. Leaf-blades linear to linear-elliptic, 0.7–2.5(–3.5) cm. long, 1–5 mm. wide, acute at the apex, cuneate at the base, ± revolute, adpressed strigose with white hairs on both surfaces; petiole ± 2 mm. long. Inflorescence spike-like of axillary flowers with the leaves mostly reduced to bracts or absent, 2.5–7.5 cm. long. Calyx-lobes unequal, lanceolate, 1.5 mm. long, 0.5–0.8 mm. wide, obtuse to subacute, glabrous to spreading hairy or adpressed pubescent outside, glabrous inside, often ± connivent in fruit. Corolla white, greenish or yellow-green; tube 1.5 mm. long, adpressed to spreading pubescent above, upper half hairy inside; lobes ovate, 1 mm. long and wide, adpressed hairy outside. Ovary glabrous; style 2–3 times longer than the stigma, the two together ± 1 mm. long; stigma capitate, globose or subconic, with or without a short apical projection but with an apical tuft of hairs. Nutlets ovoid-segment-shaped, 1 mm. tall, 1.5 mm. wide, the contiguous walls hollowed, adpressed to spreading pubescent.

subsp. **rariflorum**

Nutlets with spreading bristly hairs.

NOTE. See note.

subsp. **hereroense** (*Schinz*) *Verdc.*, comb. et stat. nov. Type: Namibia, Hereroland, *Lüderitz* 7 (B, holo.†)

Nutlets with adpressed strigose indumentum. Fig. 19/1.

KENYA. Northern Frontier Province: Dandu, 2 May 1952, *Gillett* 13008!; Masai District: Ol Lorgosailie [Orgasaile], 11 Aug. 1951, *Verdcourt* 582!; Teita District: Tsavo National Park East, Voi Gate to Sobo road, km. 35, 20 Dec. 1966, *Greenway & Kanuri* 12814!
TANZANIA. Masai District: Engaruka road, 24 Feb. 1970, *Richards* 25508!; Lushoto District: 3 km. WSW. of Mkomazi on Moshi road, 28 Mar. 1975, *Hooper & Townsend* 1035!; Kilosa District: Ruaha Gorge, about 35 km. along Mbuyuni–Mikumi road, 29 Dec. 1971, *Wingfield* 1816!
DISTR. **K** 1, 2, 4, 6, 7; **T** 2, 3, 6; Sudan (*fide* F.P.S.), Ethiopia, Somalia, Angola and Namibia
HAB. Grassland and bushland in dry areas, mainly *Combretum-Commiphora-Lannea*, *Commiphora-Dobera* and *Commiphora-Acacia* associations, often the dominant or codominant undershrub; in T6 on sand-banks subject to submergence; 100–1350 m.

SYN. *H. sp. aff. H. cordofanum* DC. & A.DC. sensu Vatke in Linnaea 43: 317 (1882), quoad *Hildebrandt* 2587
[*H. strigosum* sensu Bak. & Wright in F.T.A. 4(2): 41 (1905), quoad *Hildebrandt* 2587, *non* Willd.]
H. hereroense Schinz in Viert. Nat. Ges. Zürich 60: 404 (1915); M. Holzh. in Prodr. Fl. SW.-Afr. 119: 7 (1967)
H. pseudostrigosum Dinter in F.R. 18: 250 (1922). Types: Namibia, Dorstrivier, *Dinter* 159 & Seskamelboom, *Dinter* 2062 & Marienthal, *Dinter* (? SAM, syn.)

NOTE. Said to be an important constituent of dry scanty grazing areas and eaten by cattle. Despite the fact this is a well-defined species there are certainly some specimens where the stigma is very similar to that of *H. strigosum* even if the habit is not. Whether this is due to hybridization or variation is not yet clear; e.g. *Carter & Stannard* 699 (Kenya, Northern Frontier Province, Kaisut

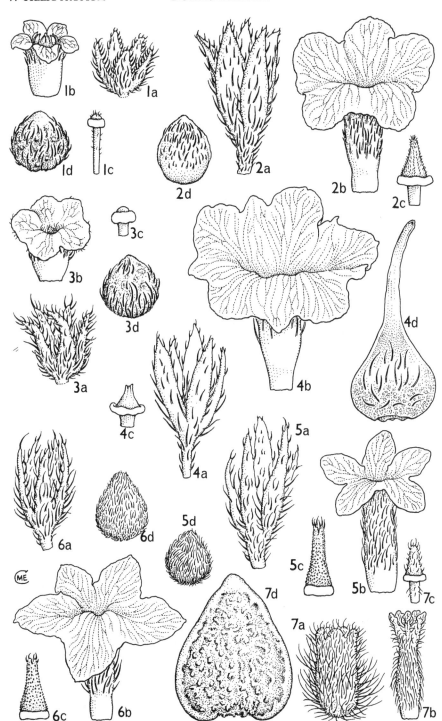

FIG. 19. *HELIOTROPIUM SPP.*—**a**, calyx, × 10; **b**, corolla, × 10; **c**, style and stigma, × 20; **d**, nutlet, × 12 of **1**, *H. RARIFLORUM* subsp. *HEREROENSE*; **2**, *H. STRIGOSUM*; **3**, *H. SESSILISTIGMA*; **4**, *H. BACLEI* var. *ROSTRATUM*; **5**, *H. BULLOCKII*; **6**, *H. OVALIFOLIUM*; **7**, *H. SUPINUM*. 1, from *Stannard & Gilbert* 1085; 2a–c, from *Drummond & Hemsley* 3778; 2d, from *Richards* 23429; 3a–c, from *Greenway* 9569; 3d, from *Gilbert & Thulin* 1380; 4, from *Verdcourt* 3447; 5, from *Bullock* 2284; 6, from *Richards* 8439; 7, from *Polhill & Greenway* 447. Drawn by Mrs Maureen Church.

Desert, 45 km. N. of Laisamis, 25 Nov. 1977) has the habit of *H. rariflorum* including the peeling silvery epidermis but the style has a distinct apical projection beyond the stigma, although the style is longer than in true *H. strigosum*. A few specimens have almost entirely glabrous flowers, e.g. *Bally & Smith* 14495! (Kenya, Northern Frontier Province, 80 km. SW. of Wajir, 10 Dec. 1951). The plant is attacked by an insect which produces silvery ellipsoid galls but it is not yet known what causes them. Subsp. *rariflorum* with spreading pubescent nutlets occurs in Arabia, Pakistan, Baluchistan, etc., and something very similar in Mauritania.

A specimen from '**K**1/3 Bal Lak', 15 Dec. 1982, *Powys* G22, described as a small herb growing low on the ground has the style and nutlet indumentum of subsp. *rariflorum* but the habit and foliage of *H. strigosum*. I hesitate to give new names to variants in this confused group without further study. The habitat is given as "lava rock on side of lugga".

15. **H. strigosum** *Willd.*, Sp. Pl., ed. 4, 1: 743 (1798); Boiss., Fl. Orient. 4: 143 (1875); C.B. Cl. in Fl. Brit. Ind. 4: 131 (1883); Hiern, Cat. Afr. Pl. Welw. 1: 719 (1898); Bak. & Wright in F.T.A. 4(2): 41 (1905); F.P.N.A. 2: 129 (1947); F.P.S. 2: 87 (1956); E.P.A.: 777 (1962); Heine in F.W.T.A., ed. 2, 2: 322 (1963); Kazmi in Journ. Arn. Arb. 51: 153 (1970); Taton, F.C.B., Borag.: 34, map 11 (1971); Kabuye & Agnew in U.K.W.F.: 518 (1974); Hepper, W.A. Herb. Isert & Thonning: 32 (1976); Vollesen in Opera Bot. 59: 76 (1980); Martins in F.Z. 7(4): 85, t. 27 (1990). Type: Ghana [Guinea], *Isert* (B, holo., C. iso.!, microfiche 219/3253!)

Typically in W. Africa annual* but elsewhere often (or almost entirely in parts of Asia) perennial with woody taproot; spreading or erect herb 7–45 cm. long or tall, sparsely to densely branched, the strictly erect unbranched to little-branched annuals appearing very different from much-branched straggling perennials. Young stems strigose with adpressed upwardly directed white hairs, later glabrescent and with peeling brown epidermis and bark. Leaves laxly to very densely spaced; blades linear, linear-lanceolate, oblong or very narrowly elliptic, 0.4–4 cm. long, 0.7–5 mm. wide, acute at apex, cuneate at the base, sparsely to densely silvery strigose on both surfaces, the margins ± revolute; petiole 0.5–2 mm. long. Flowers basically axillary but typically the very reduced bract-like leaves result in a spike-like cyme. Cymes very numerous, simple or less often branched, unilateral, 1–7 cm. long; bracts similar to the leaves in shape, 1–3 × 0.4 mm., ciliate and pubescent. Flowers ± laxly dispersed on the strigose rhachis; pedicels 0.2–0.4 mm. long, accrescent to 0.5–2(–5) mm. in fruit, angular and decurrent. Calyx 2–3 mm. long, scarcely accrescent in fruit; lobes oblong-ovate or narrowly elliptic, 1.5–2.5 mm. long, 0.3–1.2 mm. wide. Corolla white or greenish white, often with a yellow centre, funnel-shaped, 2.5–3(–5) mm. long; tube enlarged about the middle, strigose outside save at the base, inside with short flat hairs near tops of anthers; lobes broadly triangular or ovate, 0.5–2 mm. long, 0.7–2 mm. wide, the limb 3–5.5 mm. wide. Style with lower part 0.2–0.5 mm. long and narrowly conical ± papillate apical part 0.3–0.8 mm. long, either longer than or ± equalling the stipe and with a basal stigmatic disc, often minutely bifid at the apex; or with upper part above disc as narrow as part beneath, the whole resembling a wheel on an axle. Fruits globose, depressed, (1.8–)2.5–3 mm. diameter, adpressed pubescent, breaking into 4 nutlets ± 1.5 mm. wide, the internal faces with a ± central circular depression. Fig. 19/2, p. 71.

UGANDA. W. Nile District: hill SE. of Metu Rest Camp, 15 Sept. 1953, *Chancellor* 259!; Karamoja District: Kotido, Oct. 1958, *J. Wilson* 571!; Busoga District: 1.6 km. N. of Mugindi Hill, 23 May 1951, *G.H. Wood* 201!
KENYA. Northern Frontier Province: Mandera, War Gedud, 1 May 1978, *Gilbert & Thulin* 1271!; 48 km. S. of Embu, 29 May 1956, *Bogdan* 4187!**; Masai District : 8 km. S. of Magadi, 28 Mar. 1959, *Greenway* 9570!
TANZANIA. Masai/Mbulu Districts: Lake Manyara, outside National Park, 2 May 1965, *Richards* 20322!; Uzaramo District: Dar es Salaam, University College, 15 Mar. 1968, *Mwasumbi* 10333!; Iringa District: Msembe, Ruaha National Park Headquarters, 22 Feb. 1970, *Greenway & Kanuri* 13923!
DISTR. **U** 1–3; **K** 1–7; **T** 1–8; Mauritania and Senegal to Nigeria, Sudan, Ethiopia, Djibouti, Zambia, Zimbabwe, Malawi, Mozambique, Botswana, Angola, Egypt, Arabia, Afghanistan to tropical Asia, China and Australia
HAB. Grassland, *Acacia-Compositae* bushland, *Combretum* and *Brachystegia* woodlands, also *Euphorbia-Commiphora* scrub, often on bare ground, e.g. by roadsides and on open lava plains; 0–1670 m.

---

* F.W.T.A. states perennial but well over half the W. African material I have examined including the type consists of annuals.
** This is typical *H. strigosum* very like the Ghana type, with a long corolla and erect annual habit.

SYN. *H. bicolor* DC. & A. DC. in DC., Prodr. 9: 546 (1845). Type: Saudi Arabia, 'Arabia Petraea', Mt. Sidr (Sedder), *Schimper* 848 (G, syn., K, isosyn.!) & Sudan, Kordofan, *Kotschy* 208 (G, syn., BM, K, isosyn.!)*
  H. cordofanum DC. & A. DC. in DC., Prodr. 9: 546 (1845). Types: Sudan, Kordofan, Arasch-Cool, *Kotschy* 116 (G, syn., BM, K, isosyn.!) & without exact locality, *Kotschy* 96 (G, syn., BM, K, isosyn.!)
  H. longifolium Klotzsch in Peters, Reise Mossamb., Bot. : 251 (1861). Type: Mozambique, Rios de Sena, *Peters* (B, holo.†)
  H. pygmaeum Klotzsch in Peters, Reise Mossamb., Bot. : 252 (1861). Type: Mozambique, near Tete, *Peters* (B, holo.†)
  H. senense Klotzsch in Peters, Reise Mossamb., Bot. : 253 (1861). Type: Mozambique, Rios de Sena, *Peters* (B, holo.†)
  H. strigosum Willd. var. *bicolor* (DC. & A.DC.) Schwartz in Mitt. Inst. Bot. Hamburg 10: 209 (1939) (a possible earlier combination by Engler in 1910 (see Cufodontis, E.P.A.: 777 (1962)) has not been traced)

NOTE. *H. strigosum* is extremely variable even in W. Africa where the added complication of hybridization with related species is not evident. Absolutely typical specimens are common in E. Africa, particularly in coastal areas. Populations of short-leaved prostrate plants very similar to *H. sessilistigma* in habit and corolla length but with the style and stigma of *H. strigosum* may be hybrids between the two; this introgression is confined to more arid areas. A few specimens have the stigma of *H. strigosum* but the basal style longer and upper part shorter and seem somewhat intermediate with *H. rariflorum*, e.g. *Faden & Evans* 71/333 (Kenya, E. side of Lake Turkana [Rudolf], Koobi Fora to Shin Hills, 1 May 1971). *Blakenship* 10 (Northern Frontier Province, E. side of Lake Turkana [Rudolf], 7 Feb. 1970) and *Gilbert* 4744 (Baringo District, 6 km. N. of Kampi ya Samaki, 13 June 1977) have the inflorescences with very short not leafy bracts and a corolla somewhat resembling *H. rariflorum* but the stigma is that of *H. strigosum*; the peeling epidermis in the former specimen is also similar to that of *H. rariflorum*. *Gillett & Gachathi* 20568 (Northern Frontier Province, Garissa–El Lein road, 13 km. from fork off Hagadera [Haghadera] road, 10 May 1974) is also similar in floral characteristics but has very slender stems. These specimens may represent a distinct taxon for which the epithet *bicolor* might be used at some level.
  Three other variants may be distinct taxa but material is inadequate at present; two are from the Wajir area. "Wajir A" is a perennial with dense branching and linear leaves, peeling bark rather as in *H. rariflorum*; inflorescence long and ebracteate, sepals very small under 1.5 mm. long, corolla ± 1.5 mm. long and fruit under 2 mm. wide; the style and stigma are, however, ± as in *H. strigosum*, e.g. Northern Frontier Province, Wajir area, 26 km. NE. of Habaswein, 27 Apr. 1978, *Gilbert & Thulin* 1121!, in *Commiphora-Acacia-Cordia* scrub at 230–310 m. "Wajir B"** has an exactly similar prostrate habit to *H. sessilistigma* but is annual; the style and stigma are, however, ± as in *H. strigosum* and the nutlet indumentum adpressed. Examples are *Kirrika* 67! (Wajir, 22 June 1951), *Gillett & Gachathi* 20672! (36 km. N. of Wajir on Tarbaj Road, 14 May 1974 at 330 m.) and *Gillett* 21296! (8 km. E. of Wajir, 1 June 1977, at about 600 m.), *A. Curtis* R28 (Kitui/Kilifi District, Galana Ranch, 5 Oct. 1971) is probably the same. The third variant is also prostrate with numerous graceful stems from the rootstock and resembles *H. baclei* in habit; the leaves are narrowly elliptic, 10 × 3 mm., dark green and ± sparsely strigose; the sepals attain 3 × 1 mm. — *Stannard & Gilbert* 1076! (Kenya, Tana River District, 22 km. from Garissa towards Thika, 15 Dec. 1977). Kazmi keeps up a subsp. *brevifolium* (Wall.) Kazmi (Journ. Arn. Arb. 51: 153 (1970); types: India, cult. Carey (ubi? syn.) & Nepal, Katmandu & Gosain-Than (syn. — K-WALL 914.2! is one of these presumably)).

16. **H. sessilistigma** *Hutch. & E.A. Bruce* in K.B. 1941: 160 (1941); E.P.A.: 776 (1962); Kabuye & Agnew in U.K.W.F.: 518 (1974). Type: N. Somalia, boundary at 44°15′ E., *Gillett* 4107 (K, holo.!)

Densely branched intricate heath-like shrublet or more often a perennial prostrate mat-forming or ascending subshrubby much-branched herb 7–45 cm. tall; stems densely covered with ascending ± adpressed white hairs when young, later often with white peeling epidermis revealing brown bark beneath (as in *H. rariflorum*). Leaves silvery or grey-green; blades narrowly oblong-elliptic to distinctly elliptic, 0.5–1.5 cm. long, 1.5–5 mm. wide, acute at the apex, cuneate at the base, strigose with adpressed white hairs on both surfaces; petiole 0.5–2 mm. long. Flowers sessile, basically axillary but leaves often reduced so that inflorescences appear to be spike-like, ± 5 cm. long, with some flowers supported by leaf-like bracts and others not. Calyx-lobes linear to lanceolate, 1.2–2 mm. long, 0.2–0.5 mm. wide, densely spreading or adpressed pilose outside. Corolla white, scarcely exceeding the calyx; tube 1–2 mm. long, adpressed hairy outside above, glabrous beneath, hairy inside at base of lobes; limb 2–3 mm. wide, the lobes broadly ovate, rounded or scarcely defined, 0.8–1 mm. long and wide. Stigma globose or with a very short ± imperceptible projection or small tuft of hairs, the style very short or ± obsolete, the two

* A. DC. adds a footnote 'et in Cordofan, *Kotschy* 208'.
** See addendum on p. 119.

together 0.3–0.5 mm. long and often completely immersed in the central cavity between the 4 nutlets. Fruit depressed globose, ± 3 mm. wide, easily separating into 4 globose or ovoid-segment-shaped nutlets 1.2–1.5 mm. long, 1–1.2 mm. wide, typically with sparse to dense long spreading white hairs often ± dark at tips or hairs sometimes ½-adpressed. Fig. 19/3, p. 71.

KENYA. Northern Frontier Province: 40 km. on the El Wak–Wajir road, 29 Apr. 1978, *Gilbert & Thulin* 1194!; Turkana District: 128 km. N. of Lodwar (near Lokitaung), 21 May 1953, *Padwa* 189!; Masai District: 7.2 km. Ol Orgasailie–Magadi, 4 Apr. 1969, *Napper, Greenway & Kanuri* 1992! & same area, 97.6 km. Nairobi–Magadi road, 27 Mar. 1959, *Greenway* 9569!

DISTR. **K** 1, 2, 6, 7; Ethiopia, Somalia

HAB. Lava desert with scattered trees and scrub, old alkaline lake deposits, rocky cliffs and limestone pavements, bushland with *Balanites, Acacia, Commiphora*, etc.; 300–1050 m.

NOTE. *Padwa* 189, cited above, an ascending herb from a simple slender woody rootstock, was at first thought to be a distinct taxon but the floral characters are exactly as in typical *H. sessilistigma*. As pointed out under *H. strigosum* many specimens of that species have exactly the habit of *H. sessilistigma* but are easily distinguished by the style being distinctly produced beyond the stigmatic disc and adpressed nutlet indumentum; some, however, all from Kenya, have the projection much reduced and the disc more globose being genuinely intermediate and perhaps hybrids, e.g. *Gilbert et al.* 5078 (40 km. N. of Rumuruti, 25 Oct. 1978), *Le Pelley* in A.D. 3504 (Nairobi Dam, 25 June 1950) and *Bogdan* 842 (Thika, 30 June 1947) but all these have adpressed nutlet indumentum.

There is no doubt that *H. rariflorum, H. strigosum* and *H. sessilistigma* belong to a difficult complex with difficult intermediates but the majority of specimens are easily named despite intrinsic variation.

17. **H. baclei** *DC. & A.DC.*, Prodr. 9: 546 (1845); Bak. & Wright in F.T.A. 4 (2): 34 (1905); I.M. Johnston in Contr. Gray Herb. 92: 91 (1930); Heine in F.W.T.A., ed. 2, 2: 322 (1963). Type: Gambia, Quoja [Quoia], *Bacle* (G, holo.)

Perennial herb, woody at the base, forming a rosette or mat with prostrate or suberect branched stems 7–40 cm. long, densely strigose with white adpressed hairs or sometimes glabrescent. Leaves alternate, elliptic, oblong or oblanceolate, 0.4–1.7 cm. long, 1.5–6(–8) mm. wide, acute at the apex, cuneate at the base, entire, strigose-pubescent with adpressed tubercle-based hairs each surrounded by a zone of cystoliths. Flowers solitary, supra-axillary at the ends of leafy branches; pedicels 0.2–0.3 mm. long, pubescent. Calyx funnel-shaped, divided to near the base; lobes lanceolate, subequal, 2.8–4.2 mm. long, 0.7–0.9 mm. wide, or unequal with 1 or sometimes 2 much larger 5–5.5 mm. long, 1.2–1.6 mm. wide, acute, sparsely strigose outside, ciliate at the base. Corolla yellow or white with a yellow throat, funnel-shaped or subrotate, sparsely pubescent outside, glabrous inside; tube 3–5 mm. long; lobes broadly triangular, 1–1.5 mm. long, 1–1.2 mm. wide, rounded at the apex, sometimes alternating with minute teeth. Anthers subsessile, coherent around the stigma. Ovary ovoid, 0.7–0.8 mm. diameter, glabrous or sparsely pubescent, surrounded by a cupuliform disc; style 0.2–0.4 mm. long; stigma conical with a plate-like stigmatic ring at base. Fruit ovoid, 3.6–5 mm. long, 2 mm. diameter, with a beak 0.5–3(–5) mm. long, sparsely pubescent with long hairs, dividing into 4 nutlets; pedicel accrescent to 2 mm.

var. **rostratum** *I.M. Johnston* in Contr. Gray Herb. 92: 91 (1930); Heine in F.W.T.A., ed. 2, 2: 321 (1963); Taton, F.C.B., Borag.: 35, map 11 (1971); Vollesen in Opera Bot. 59: 76 (1980); Martins in F.Z. 7(4): 83, t. 26/1 (1990). Types: Zaire, Shaba, Lukafu, *Verdick* 141 & 182 (BR, syn.)

Corolla mostly entirely bright yellow. Fruits with beak to 3 mm. long (5 *fide* Heine). Fig.19/4. p. 71.

TANZANIA. Buha District: 64 km. on Kibondo–Kasulu road, Malagarasi Ferry, 24 Nov. 1962, *Verdcourt* 3447!; Mpanda District: Lake Katavi [Kitavi], 29 Nov. 1956, *Richards* 7080!; Mbeya District: Great Ruaha R. near the Trekimboga Hippo Pools, 12 Dec. 1970, *Greenway & Kanuri* 14787!

DISTR. **T** 4, 6, 7; Mali, Central African Republic, Zaire, Ethiopia, Malawi, Zambia, Zimbabwe, Caprivi Strip, Botswana, Angola and Namibia

HAB. Muddy and sandy river-edges and lake-shores, bare or with grasses, Cyperaceae and other ephemeral sand-bank vegetation; (300–)780–1200 m.

SYN. *H. katangense* De Wild. in Ann. Mus. Congo, Bot., sér. 4, 1: 223 (1903); Bak. & Wright in F.T.A. 4(2): 43 (1905); De Wild., Contr. Fl. Katanga: 162 (1921). Type as for var. *rostratum* [*H. marifolium* sensu Bak. & Wright in F.T.A. 4(2): 40 (1905), pro parte, *non* Retz.]

NOTE. Var. *baclei* occurs from Senegal to Sierra Leone.

18. **H. bullockii** *Verdc.*, sp. nov. in subgenere *Orthostachye* (R.Br.) Riedl ponenda a propinquis africanis, stylo ± obsoleto, cono stigmatifero papillato apice piloso, foliis usque 7 × 1.5 cm. sparse strigosis, lobis calycis inaequalibus differt. Typus: Tanzania, Mpanda District, Katisunga, *Bullock* 2284 (K, holo.!)

Herb at least 40 cm. tall but habit not known, probably straggling and perennial; stems somewhat woody at the base, with upwardly directed white adpressed hairs in no way obscuring the brown epidermis. Leaf-blades oblanceolate, narrowly oblong or narrowly elliptic, 2–7.5 cm. long, 0.3–1.5 cm. wide, abruptly acute at the apex, narrowly cuneate into the petiole, with adpressed ± tubercle-based upwardly directed hairs on both surfaces, sometimes a little spreading on the 1.2 cm. long petioles. Inflorescences forked, each scorpioid cyme 9 cm. long, the open flowers crowded at apex but after flowers fall the immature fruits are well-spaced with internodes 0.5–1 cm. long; peduncle 6 cm. long, together with axes with similar indumentum to the stems. Calyx-lobes very unequal, all bristly white hairy, the largest 2.6 mm. long, 0.8 mm. wide, others 2.2 × 0.6, 2 × 0.5 and the smallest linear, 1.8 × 0.2 mm.; tube ± 0.5 mm. long. Corolla white, tube ± 3 mm. long, constricted below the throat, glabrous at base outside, adpressed white-pubescent above; lobes oblong-elliptic, 1.2 mm. long, 0.8 mm. wide. Anthers situated about the middle of the tube, 1.3 mm. long including narrow apical appendage. Ovary 0.5 mm. long; stigma sessile, narrowly conical, 1 mm. long, minutely papillate with a few hairs at the apex. Fruits not seen. Fig.19/5, p. 71.

TANZANIA. Mpanda District: Katisunga, 20 Jan. 1950, *Bullock* 2284!
DISTR. T 4; not known elsewhere
HAB. Bare soil in hollows of mbuga; 1050 m.

19. **H.ovalifolium** *Forssk.*, Fl. Aegypt.-Arab. : 38 (1775); C.B.Cl. in Fl. Brit. Ind. 4: 150 (1883); Gürke in P.O.A. C: 337 (1895); Hiern, Cat. Afr. Pl. Welw. 1: 718 (1898); Wright in Fl. Cap. 4(2): 8 (1904); Bak. & Wright in F.T.A. 4(2): 34 (1905) (synonymy); Z.A.E. 2: 280 (1911); F.P.N.A. 2: 129 (1947); I.M. Johnston in Journ. Arn. Arb. 32: 111 (1951); E.P.A.: 775 (1962); Heine in F.W.T.A., ed. 2, 2: 322 (1963); Kazmi in Journ. Arn. Arb. 51: 178 (1970); Taton, in F.C.B., Borag.: 32, map 10 (1971); Kabuye & Agnew in U.K.W.F.: 520, fig. on 519 (1974); Vollesen in Opera Bot. 59: 76 (1980); Martins in F.Z. 7(4): 87, t. 26/2 (1990). Type: N. Yemen, Al Hadiyah [Hadïe], *Forsskål* 299 (C, holo., photo.!, BM, iso.)

Usually a perennial herb 7–90 cm. long or tall with thick woody taproot and stems branched, procumbent, woody at the base but sometimes erect and apparently sometimes annual (?flowering in first year); stems often compressed, very leafy, silky-silvery pubescent on the young branches. Leaf-blades obovate or elliptic, 0.5–5.5 cm. long, 0.25–2.5 cm. wide, retuse but mucronate or apiculate or less often acute at the apex, cuneate at the base, pubescent to strigose-villous, often silvery on both surfaces, the hairs of two sorts, longer and tubercle-based and shorter and thinner, the margins entire, narrowly thickened, only the costa distinctly visible; petiole up to 1.5(–2) cm. long. Cymes 1–3, terminal, scorpioid, spike-like, often paired, at first often very short and ± capitate, 1.5–15 cm. long, strigose; flowers disposed in 2 ranks, at first subsessile; peduncles up to 3.5 cm. long; pedicels accrescent, 0.5–2 mm. long finally decurrent. Calyx 1.6–3.5 mm. long, strigose outside, persistent and accrescent in fruit, deeply divided into unequal lanceolate lobes, 1.5–3.3 mm. long, 0.3–1.2 mm. wide. Corolla white, often yellow at the throat, strigose-pubescent outside with adpressed hairs; tube cylindrical, enlarged at the middle, (1.4–)2.3–3.7 mm. long, with a ring of adpressed hairs inside at tops of anthers; lobes broadly elliptic to ovate-triangular, erect or spreading, 0.7–1 mm. long, 0.4–1.5 mm. wide, the acute apices sometimes inflexed. Ovary subglobose, 0.5 mm. diameter, glabrous or pubescent at the tip. Stigma sessile, conic with basal crown, 0.5–0.7 mm. long, truncate or sometimes very shortly bifid with several erect hairs at apex. Fruits subglobose, flattened, 1.7–2 mm. diameter, 4-lobed and breaking into 4 nutlets, the external face densely covered with short ± adpressed white hairs; internal faces devoid of cavities. Fig. 19/6, p. 71.

KENYA. Kitui District: Athi R., N. of the Kibwezi–Kitui low-level bridge, 22 Apr. 1969, *Napper & Kanuri* 2044!; Masai District: Uaso [Ewaso]-Nyiro R. where road from Lake Magadi bridges it, 21 Jan. 1951, *Verdcourt, Greenway & Eggeling* 421!; Lamu District: near Mkokoni, Mar. 1957, *Rawlins* 390!
TANZANIA. Mwanza District: Ujashi, 2 Feb. 1953, *Tanner* 1187!; Mpanda District: Lake Tanganyika, Kibwesa Point, 7 Sept. 1958, *Newbould & Jefford* 2392!; Morogoro District: 3.2 km. S. of Mkata Station, 17 Jan. 1957, *Welch* 252!; Pemba I., Vitongoge, 22 Sept. 1929, *Vaughan* 670!

DISTR. **K** 1, 4, 6, 7; **T** 1, 2, 4–8; **P**; Senegal to Ethiopia and Somalia to South Africa and Angola, Madagascar, Egypt, Arabia, India, Pakistan, also introduced into Pacific and Taiwan

HAB. Essentially a plant of wet places in dry bushland, etc., e.g. on sand in *A. seyal* woodland near water, sandy beaches by dry country rivers, dried mud around pools in grassland, salt pans, also on black cotton soil near irrigation channels; ± 0–1635 m.

SYN. *H. phyllosepalum* Bak. in K.B. 1894: 30 (1894); Bak. & Wright in F.T.A. 4(2): 33 (1905). Type: Mozambique, banks of Shire R. at Moramballa, *Scott* (K, holo.!)

NOTE. A specimen from 'am Albert Edward See' *Mildbraed* 1906 was I think collected in Zaire but it seems likely it may be found in Uganda nearby.

Some weakly erect annual plants with long petiolate much greener leaves and very short capitate inflorescences may be this species flowering in its first year or as an ephemeral but they are very distinct in general appearance e.g. *Mathew & Gwynne* 6763 (Northern Frontier Province, S. Turkana, Ayangiyangi Swamp, 12 June 1970). A number of other specimens appear to be this species but are atypical in having much less silvery leaves. *Vaughan* 670 cited above is an erect annual; it had been referred to *H. phyllosepalum*. I do not think var. *depressum* (Cham.) Merr. (type: Pacific, Guam, *Chamisso* (LE, holo., K, iso.!)) is separable. Exactly similar specimens occur in Tanzania. It must have been introduced into the Pacific at an early date. Much irrelevant synonymy from other areas has been omitted. *H. barbatum* DC. & A.DC., described from Brazil, Bahia is clearly very close to *H. ovalifolium* but only two old specimens have been seen.

20. **H. supinum** *L.*, Sp. Pl.: 130 (1753); Sibthorp & Smith, Fl. Graec. 2, t. 157 (1813); DC. & A.DC. in DC., Prodr. 9: 533 (1845); C.B.Cl. in Fl. Brit. Ind. 4: 149 (1883); Engl., Hochgebirgsfl. Trop. Afr.: 351 (1892); Hiern, Cat. Afr. Pl. Welw. 1: 717 (1898); Wright in Fl. Cap. 4(2): 8 (1904); Bak. & Wright in F.T.A. 4(2): 37 (1905); Gamble, Fl. Madras 2: 896 (1923); Bonnier, Fl. Compl. Fr., Suisse et Belg. 8: 19, t. 429/2012 (1926); Heine in F.W.T.A., ed. 2, 2: 322 (1963); Riedl in Fl. Iran. 48/15: 52 (1967); Kazmi in Journ. Arn. Arb. 51: 179 (1970); Täckh., Students' Fl. Egypt, ed. 2: 442, t. 153/A (1974); Kabuye & Agnew in U.K.W.F.: 520 (1974); Riedl in Fl. Turkey 6: 255, fig. 11/12 (1978); Meikle, Fl. Cyprus: 1123 (1985); Verdc. in K.B. 42: 710 (1987); Martins in F.Z. 7(4): 93, t. 26/11 (1990). Type: France, seashore near Montpellier, specimen in Burser Herbarium Vol. XIV (2): 2 (UPS, lecto.)*

Decumbent or procumbent annual 4–50 cm. long or tall or forming mats up to 30 cm. in diameter; stems densely silvery pubescent and spreading pilose and similar dual indumentum on most organs. Leaf-blades grey-green, elliptic, ovate, obovate or ovate-lanceolate, 0.8–3.5 cm. long, 0.6–1.5 cm. wide, rounded to acuminate at the apex, cuneate at the base, plicate, the nerves deeply impressed above, densely white adpressed pubescent above with tubercle-based hairs and longer setae, with more tangled soft white pubescence beneath and longer hairs which are often brownish; petiole 0.2–2 cm. long. Flowers sessile in bilateral scorpioid cymes 2–9 cm. long; peduncles 0–3 cm. long. Calyx 2–2.5 mm. long, divided into linear densely spreading white-pilose subacute or ± obtuse lobes. Corolla white, 2.5 mm. long; tube cylindric, ± 3 mm. long, pilose outside, glabrous inside; limb very small, with 5 ovate spreading rounded imbricate lobes. Stigma elongate-conic, hairy at the apex and about equalling the glabrous style, together ± 1 mm. long. Fruits of 1–4 nutlets or 1 only by abortion, remaining in the accrescent calyx with the corolla still attached at the apex; in Europe the nutlets are almost invariably reduced to 1, compressed ovoid, (3–)3.5–4 mm. long, 2.8–3.2 mm. wide, 1.5 mm. thick, plano-convex or biconvex, acute at the apex, the outer face smooth to verrucose, distinctly margined with obtuse compressed edge; in the tropics there are usually 2–4 (mainly 4) nutlets cohering but ultimately or sometimes readily separating, often rather narrower than are the single nutlets, similarly smooth to verrucose; eventually the nutlets are often smooth or ± corky. Fig. 19/7, p. 71.

UGANDA. Karamoja District: Bokora County, July 1957, *J. Wilson* 384!

KENYA. Northern Frontier Province: S. Turkana, Ayangiyangi Swamp, 12 June 1970, *Mathew & Gwynne* 6776!; Machakos District: Stony Athi, 7 Apr. 1940 (fl.), 6 Oct. 1940 (fr.), *E.A. Nat. Hist. Soc.* 148!; Masai District: below Ol Orgasailie, Ol Keju Nyiro [Ol Kejo Neru] R., 25 Aug. 1963, *Verdcourt* 3716!

TANZANIA. Kwimba District: 45 km. from Mwanza on Musoma road, 17 July 1960, *Verdcourt* 2896!; Dodoma District: Mwitikira, 16 Aug. 1928, *Greenway* 776!; Iringa District: by Great Ruaha R., on Great North Road, 17 July 1956, *Milne-Redhead & Taylor* 11236!

DISTR. **U** 1; **K** 1, 2, 4, 6, 7; **T** 1–5, 8; Egypt, N. Africa, Canary Is., Senegal, Mali, Niger, Nigeria, Cameroon, Mozambique, Zimbabwe, Zambia, Botswana, Angola, South Africa and Namibia; also S. Europe, Palestine, Syria, Arabia, Iraq, Iran and India

* Riedl gives the type as 'hab. Salmanticae juxta agros', *Alstroemer* 482 (Herb. Linn. 179/8) but this specimen was not in Linnaeus' herbarium until 1762 and cannot be the type; Clusius (Rariorum Plantarum Historia lib. 4, xlvii (1601)) cited by Linnaeus mentions Salamanca but not Montpellier.

HAB.   Essentially on dried up riverine soil; lake-shores, dried river beds, very often on alluvial or black clay soils even when hard like concrete, mud under *Acacia seyal* and in *Cordia, Capparidaceae, Salvadora, Pluchea* associations, dried up ponds etc. and sometimes remaining as a weed during cultivation; 375–1500 m.

SYN.   *Lithospermum heliotropioides* Forssk., Fl. Aegypt.-Arab. : 39 (1775). Type: Egypt, Cairo, *Forsskål* (C-FORSSK. Herb. 302, holo., photo.!)
   Heliotropium malabaricum Retz., Obs. Bot. 4: 24 (1786); Fischer in K.B. 1932: 60 (1932). Malabar, *Koenig* (LD, lecto.)*
   Piptochlaina supina (L.) G. Don, Gen. Syst. 4: 364 (1837)
   P. malabarica (Retz.) G. Don, Gen. Syst. 4: 364 (1837)
   Heliotropium ambiguum A.DC. & DC. in DC., Prodr. 9: 533 (1845). Type: South Africa, Olifant R., near Ebenezar, *Drège* 7835 (G, holo., K!, PRE, iso.)
   H. supinum L. var. *malabaricum* (Retz.) A. DC. & DC. in DC., Prodr. 9: 533 (1845)

NOTE.   There is no doubt that there are two ± distinct populations or 'weak subspecies'. In Europe the fruit is practically invariably reduced to 1 nutlet whereas in tropical Asia and Africa there are usually 4 but sometimes 2 or 3 and occasionally 1. I had at first decided to maintain two subspecies but with some plants in our area definitely with only 1 nutlet decided against it. Those who wish to do so may call the tropical one var. *malabaricum* (*H. ambiguum* is the same taxon). The nutlets can be smooth, tuberculate or thickened and ± corky; whether these are stages of growth is not clear. It seems it may be an adaptation to dispersal by water and the biological significance of the abortion to 1 nutlet in northern latitudes would repay study. A very detailed study of the species throughout its range might reveal more evidence but is beyond the scope of a regional Flora. Riedl defines subgen. *Piptochlaina* partly by having only one fertile locule. C.B. Clarke mentions that "Bunge (Bull. Soc. Nat. Mosc. 42(1); 289 (1869)) excludes from the sect. *Piptochlaina* all the species with 4 nutlets and objects to regarding *H. malabaricum* as a var. of *H. supinum* but there is every gradation between the two forms of fruit in the Indian collections"; and the same is certainly true in Africa.

8. LITHOSPERMUM

L., Sp. Pl.: 132 (1753) & Gen. Pl., ed. 5: 64 (1754); I.M. Johnston in Journ. Arn. Arb. 33: 299–365 (1952)

Perennial or rarely annual or biennial erect or spreading herbs or subshrubs with hispid, strigose or densely hairy stems. Leaves alternate, entire, the venation usually obscure. Flowers yellow, orange or white, regular, sometimes heterostylous, in few–several-flowered simple or paired, terminal or axillary scorpioid cymes; bracts leafy, numerous, exceeding the calyx. Calyx-lobes 5, linear-cuneate or lanceolate, usually much shorter than the corolla-tube, the abaxial the largest. Corolla funnel-shaped or cylindrical, velvety outside; tube straight, glabrous or pubescent at the base inside, the throat with small velvety or glandular appendages, less often naked; lobes 5, spreading or rarely erect, imbricate, rounded, ovate, obovate or semicircular; nectaries on the internal face and at the base of the tube either as swellings, a narrow ring, lobes or small hairs. Stamens inserted near the middle of the tube, included. Ovary with 4 lobes; style gynobasic, simple, slender to rather stout, included or exserted; stigmas 2, globose, hemispherical or ellipsoid, terminal or situated just below the bilobed apex of the style. Fruit of 4 nutlets or less by abortion, ovoid or ellipsoid, mostly very shiny and porcellanous, rarely verrucose or rugose.

About 60 species, mainly in temperate regions of both hemispheres but particularly in Mexico and N. America.

L. afromontanum *Weim.* in Bot. Notis.[93]: 63, fig. 7 (1940); I.M. Johnston in Journ. Arn. Arb. 33: 348 (1952); A.V.P.: 158, 314, fig. 40, a,b (1957); Taton, F.C.B., Borag.: 57, map 20 (1971); U.K.W.F.: 522 (fig), 523 (1974); Martins in F.Z. 7(4): 108, t. 31 (1990). Type: Zimbabwe, at foot of Mt. Inyanga, *Norlindh & Weimark* 5070 (LD, holo., K, iso.!)

Perennial often straggling subshrubby herb 0.3–1.5 m. tall; stems branched above, strigose or hairy. Leaves elliptic to lanceolate, 3–5.5(–7.5) cm. long, 0.5–1.2(–2.2) cm. wide,

---

* Retzius also cites Burm., Fl. Indica: 40, t. 16/1 (1768) where a Garcin specimen also from Malabar is treated by Burman as a variety of *H. europaeum* but figured as *H. malabaricum*. Retzius does not actually mention Koenig but his description is clearly taken from a specimen; Fischer saw this Koenig specimen at Lund written up in Retzius' hand.

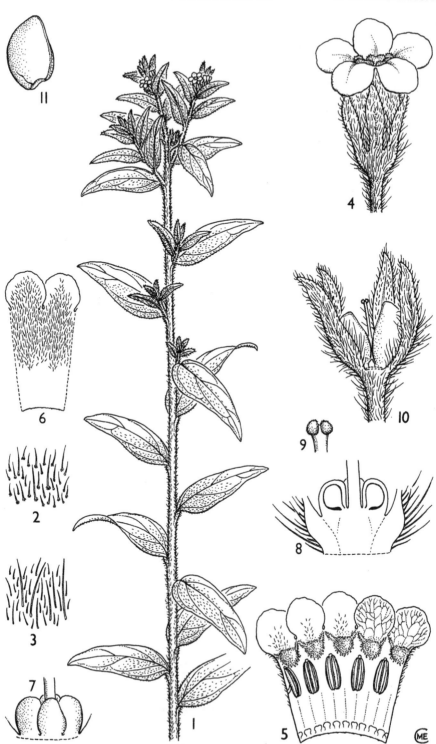

FIG. 20. *LITHOSPERMUM AFROMONTANUM*—**1**, habit, × ⅔; **2**, part of upper surface of leaf, × 16; **3**, part of lower surface of leaf, × 16; **4**, flower × 6; **5**, corolla, opened out, × 6; **6**, part of corolla from outside, × 6; **7**, ovary, × 18; **8**, ovary, longitudinal section, × 18; **9**, stigma, × 18; **10**, fruit, × 6; **11**, nutlet, × 6. 1–3, from *Greenway & Kanuri* 15048; 4–11, from *Verdcourt* 3775. Drawn by Mrs Maureen Church.

acute at the apex, narrowed to the base, discolorous, covered on both faces with hairs slightly dilated at the base, those on the upper surface often with a ring of cystolith cells; lower leaves ± sessile, the upper shortly petiolate. Cymes attaining 30 cm., many-flowered, lax; bracts subsessile, 0.5–3 cm. long, 2–9 mm. wide, ± opposed; pedicels 2–3 mm. long, accrescent to 5–9 mm. in fruit. Calyx divided to the base into often unequal lobes 3–5 mm. long, 0.6 mm. wide, accrescent in fruit and attaining 6–8 mm. Corolla white to yellow; tube 4.5–7 mm. long, pubescent outside and with papillate swellings obstructing the throat; limb 5–8 mm. wide, the lobes ± rounded, 2.2–4 mm. long and wide, ± narrowed to the base. Ovary 0.7 mm. diameter, glabrous; style 3–4(–6) mm. long with 2 minute terminal stigmas; an annular nectary inserted at the base of the tube and forming a 10-lobed swelling or with 10 completely separate lobes. Nutlets 4 or only 1 by reduction, white, ovoid, 2.8–4 mm. long, 2–3 mm. wide, smooth and shining, often with some pale brown areas when dry. Fig. 20.

UGANDA. Karamoja District: Moroto Mt., Jan. 1959, *J. Wilson* 648!; Kigezi District: Muhavura-Mgahinga saddle, Sept. 1946, *Purseglove* 2138!; Mbale District: Elgon, in crater, Jan. 1918, *Dummer* 3433!
KENYA. Elgeyo District: Cherangani Hills, Embobut Forest, Jan. 1971, *Tweedie* 3828!; N. Nyeri District: Mt. Kenya, Sirimon Track, 22 Sept. 1963, *Verdcourt* 3775!; Masai District: Narok, Nasampolai valley, 12 Aug. 1972, *Greenway & Kanuri* 15048!
TANZANIA. Moshi District: Kilimanjaro, N. slopes of Kibo, Kimengelia stream, 24 Feb. 1933, *C.G. Rogers* 494!; Mbulu/Masai Districts: Oldeani Mt. peak, 26 Nov. 1950, *Greenway* 9072!; Mbeya Mt., just N. of the peak, 13 May 1956, *Milne-Redhead & Taylor* 10335!
DISTR. U 1–3; K 2–6; T 2–4, 7; E. Zaire, Ethiopia, Malawi, Zambia, Zimbabwe and South Africa (Natal)
HAB. Montane grassland, clearings in bamboo, bushland and shrub, *Erica* and *Lycopodium* associations, *Juniper* forest; (1560–)1800–3950 m.
SYN. *L. officinale* L. var *abyssinicum* (Vatke) Engl., Hochgebirgsfl. Trop. Afr.: 355 (1892); Gürke & F. Vaupel, Z.A.E. 2: 281 (1911); Staner in Rev. Zool. Afr. 23(2): 224 (1923), nom. nud. and reference to Vatke not found
    [*L. officinale* sensu Bak. & Wright in F.T.A. 4(2): 59 (1905); F.P.S. 3:88 (1956) *non* L.]

NOTE. Long confused with and extremely similar to *L. officinale* but stems ± woody and more branched and with quite different pollen; Taton (1971) has confirmed the original work on pollen differences but it is certainly difficult to believe it is not conspecific from external characteristics. Hedberg (1957) satisfied himself about the validity of the pollen differences.
    Herlocker H-233 (Ngorongoro Conservation Area, Lemagrut Mt., 2400–3000 m., Nov. 1965) is a striking variant with spreading very closely placed narrow leaves and large flowers with corolla-tube 8–10 mm. long. It may be worth recognising, but other specimens from the Crater Highlands, e.g. the neighbouring Oldeani Mt., approach it in some ways.

9. BUGLOSSOIDES*

Moench, Meth.: 418 (1794); I.M. Johnston in Journ. Arn. Arb. 35: 38–46 (1954)

Annual setulose to strigulose herbs with simple to much-branched stems. Flowers in terminal bracteate cymes. Calyx divided almost to the base, ± strongly accrescent in fruit. Corolla white to blue or purple; throat with 5 longitudinal bands of hairs but with no scales. Filaments inserted below the middle of the tube; anthers included. Nutlets subpyriform in outline with ± incurved or almost straight beak, depressed-pyriform in dorsal view with ± bigibbous shoulders in lower half and laterally compressed beak with strong dorsal and ventral keel in upper half, tuberculate-rugose to granular.

A genus of 2–3 or 7 species (according to delimitation by Meikle or I. M. Johnston (see below)), one now widespread as a weed in many parts of the world.

B. arvensis (*L.*) *I.M. Johnston* in Journ. Arn. Arb. 35: 42 (1954); E.P.A.: 785 (1962); Fernandes in Fl. Europaea 3: 87 (1972); Edmondson in Fl. Turkey 6: 316, fig. 7/17, 10/12 (1978); Meikle, Fl. Cyprus: 1148 (1985). Type: Europe, Herb. Linn. 181/4 (LINN, syn.)

Erect or decumbent annual (2–)6–40(–90) cm. tall, with mostly simple or branched thinly to densely adpressed-strigillose or patently setulose stems. Leaves linear-oblong, lanceolate, narrowly elliptic or oblanceolate, 0.8–4.5(–10) cm. long, 0.2–1.2(–2) cm. wide,

---

* Based mostly on the accounts of Edmondson and Meikle.

FIG. 21. *BUGLOSSOIDES ARVENSIS* subsp. *ARVENSIS*—**1**, habit, × ²/₅; **2**, part of upper leaf surface, × 8; **3**, flower, × 6; **4**, corolla, opened out, × 8; **5**, ovary and style, × 30; **6**, ovary, longitudinal section, × 30; **7**, calyx with young fruit, × 3; **8**, nutlet, × 10. All from *Abdallah* 520. Drawn by Mrs Maureen Church.

acute to ± obtuse or rounded at the apex, sessile or somewhat amplexicaul at base. Cymes simple, terminal or several arising from the axils of the uppermost leaves, becoming lax and elongate in fruit; pedicels 1–3(–5 in fruit) mm. long and sometimes becoming markedly and asymmetrically incrassate in fruit. Calyx densely pale bristly; lobes erect, linear-subulate, 3–6 mm. long, becoming up to 0.8–1.5 cm. long in fruit, blunt or acute. Corolla white, pale to dark blue or purple, funnel-shaped or ± salver-shaped; tube as long as calyx or a little longer; lobes oblong, rounded or subtruncate, 1.5–2.5 mm. long, 1–2 mm. wide. Nutlets whitish or pale brown, very hard and ± glossy, 2–3(–4) mm. long, 1.5–2.5 mm. wide, minutely to coarsely rugulose-tuberculate, without lateral swellings; beak short, blunt, slightly incurved.

subsp. **arvensis**

Usually 15–40(–90) cm. tall. Pedicels not or very slightly incrassate. Calyx usually equalling or longer than corolla-tube in flower, strongly accrescent but not asymmetrical after flowering, up to 1.5 cm. long. Corolla white with base of tube often blue, 6–9.5(–16) mm. long. Nutlets usually large, 3–4 mm. long, coarsely tuberculate. Fig. 21.

TANZANIA.   Arusha District: Arumen District, Ngaramtoni, T.P.R.I. (Tropical Products Research Institute) Area, 9 Aug. 1978, *Abdallah* 520!

DISTR.   T 2; Ethiopia and South Africa; Europe, N. Africa, SW. and Central Asia, now introduced into E. Asia, Australia and N. and S. America and doubtless elsewhere

HAB.   Cultivated ground; 1380 m.

SYN.   *Lithospermum arvense* L., Sp. Pl.: 132 (1753); Boiss., Fl. Orient. 4: 216 (1875); Hegi, Ill. Fl. Mitt.-Eur. 5(3), t. 221/5 (1927); Ross-Craig, Draw. Br. Pl. 21, t. 19 (1965)

NOTE.   The solitary specimen seen from the Flora area (to which it would appear the species has only recently been introduced) has the corolla shorter than usual for the typical subspecies but cannot be referred to any of the other 3 which appear restricted to Europe and the Middle East and are sometimes treated as distinct species.

10. **ECHIUM**

L., Sp. Pl.: 139 (1753) & Gen. Pl., ed. 5: 175 (1754)

Annual, biennial or perennial hispid herbs or shrubs with an indumentum of tubercle-based setae and often an underlayer of short stiff adpressed or spreading hairs. Inflorescence thyrsoid, with lateral helicoid cymes often much enlarging in fruit. Calyx deeply 5-lobed, usually accrescent. Corolla blue, purple, yellow or white; tube broadly to narrowly funnel-shaped, straight, ± hairy outside, inside usually with a ring of 5–10 distinct scales or lobes or an undulate entire somewhat fleshy collar-like membrane at the base; throat ± oblique, open; limb usually oblique with unequal lobes. Stamens 5, often unequal, included or exserted. Style exserted; stigma bifid or sometimes capitate. Nutlets 4, ± triangular at the base, usually rugose.

About 45 species of which 27 are endemic to Madeira, Azores and Canary Is., the rest in Europe and Mediterranean region. No species is native to the Flora area but one is locally naturalised and at least one of the Canary Island species is cultivated. It is likely that several other of these striking species are also cultivated in East African gardens. Bramwell (Lagascalia 2 (1): 37–115 (1972)), has revised the Macaronesian species. The only herbarium specimen seen of this group is *Greenway* 10874 (Kenya, Kiambu District, Muguga, Hort. Greenway, 13 Mar. 1963) which appears to be the Madeiran *E. candicans* L.f. but I remember seeing similar plants frequently cultivated in the Kenya Highlands. Jex-Blake, Gard. E. Afr., ed. 4: 63, 86, 111 (1957), mentions annual *Echium* Viper's Bugloss (perhaps *E. vulgare* L., similar to *E. plantagineum* but with coarser indumentum, 4 exserted stamens and usually shorter flowers) and also *E. fastuosum, E. rubrum* and *E. strictum* as shrubs. It is not possible to know what is meant by the first name without specimens since it has been applied to several species but *E. strictum* L. is a Canary Is. species with a laxer inflorescence and wider leaves than *E. candicans; E. rubrum* Jacq. *non* Forssk., properly known as *E. russicum* Gmel., is a biennial similar to *E. vulgare* but has a ± bilobed but capitate stigma, not a bifid one, but since Jex-Blake speaks of great red spikes the true identity of this plant is not known.

E. plantagineum *L.*, Mant. Pl. Alt.: 202 (1771); Bonnier, Fl. Compl. Fr., Suisse et Belg. 7: t. 420/1979 (1924); Hegi, Ill. Fl. Mitt.-Eur. 5(3), fig. 3139 (1927); Gibbs in Lagascalia 1: 57 (1971); Edmondson in Fl. Turkey 6: 322, fig. 8 (1978); Piggin in Journ. Austral. Inst. Agric.

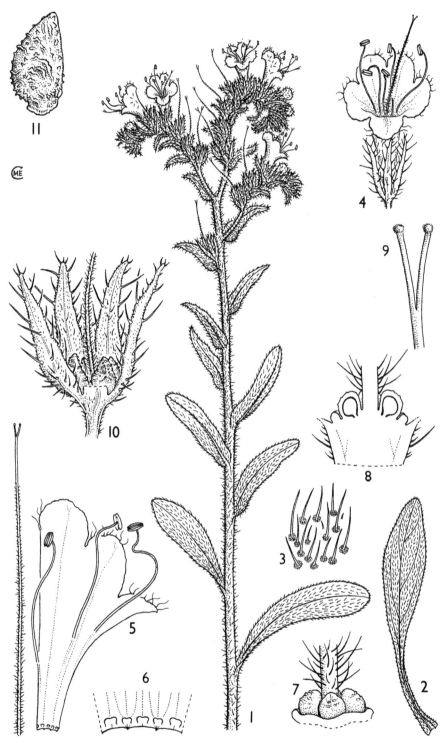

FIG. 22. *ECHIUM PLANTAGINEUM*—**1**, habit, × ²⁄₅; **2**, basal leaf, × ²⁄₅; **3**, part of upper leaf surface, × 8; **4**, flower, × 2; **5**, corolla, longitudinal section, × 3; **6**, base of corolla, × 12; **7**, ovary, × 24; **8**, same, longitudinal section, × 24; **9**, style, × 24; **10**, fruit, × 4; **11**, nutlet, × 10. 1–3, from *Thurston* s.n.; 4–11, from K783. Drawn by Mrs Maureen Church.

Sci. 48: 3–16 (1982) (biology); Meikle, Fl. Cyprus: 1155 (1985). Lectotype chosen by Gibbs (1971): Barrelier, Plantae per Galliam ... exhibitae: 145, t. 1026* (1714) (lecto.)

Erect annual to biennial herb with 1–many flowering stems 20–65 cm. tall; radical leaves ovate to oblanceolate, 5–28 cm. long, 1–3.5 cm. wide, adpressed setose with lateral nerves distinct; cauline leaves oblong to lanceolate, the upper cordate or truncate at the base, all with adpressed rather soft hairs. Cymes usually distinctly stalked, lengthening after flowering to 25–30 cm. Calyx 0.7–1.2 cm. long, lengthening to 1.7 cm. in fruit. Corolla blue turning purple then pink or entirely white, 1.8–3.2 cm. long, with hairs restricted to the veins and margins; ring of 10 distinct pilose lobes within the tube. Lower pair of stamens exserted, the others included or only slightly exserted; filaments with long woolly hairs. Style included, densely hispidulous below, bifid for ± 1 mm. Nutlets black or grey-brown, ovoid-bipyramidal, 2.3–3 × 2–2.5 mm. with prominent ventral and distinct dorsal keel and short suberect beak, tuberculate and faintly striate. Fig. 22.

KENYA. Nakuru District: Molo, Oct. 1958, *Ivens* 1261! & ? Nakuru, 30 Nov. 1965, *Dalgety Ltd.* in *E.A.H.* 13428!
TANZANIA. Lushoto District: Mkusi, 31 Aug. 1950, *Verdcourt* 340!
DISTR. **K** 3; **T** 3; Zimbabwe, Angola; native to W. Europe including SW. Britain and Mediterranean area introduced into other parts of Europe, S. Russia and Caucasus and widely distributed as a casual and a bad weed in Australia
HAB. Naturalised along roads and in meadows; 1680–2420 m.

SYN. [*E. lycopsis* sensu auctt. mult. et L., Fl. Angl.: 12 (1754), pro parte sed lectotypo excluso]

11. **CYSTOSTEMON**

Balf.f. in Proc. Roy. Soc. Edin. 12: 82 (1882) & in Trans. Roy. Soc. Edin. 31: 186, t. 56 (1888), as "*Cystistemon*"; A.G. Miller & Riedl in Notes Roy. Bot. Gard. Edin. 40: 1 (1982)

Vaupelia Brand in F.R. 13: 82 (1914), *nom. conserv.*

Canescent herbs or subshrubs, setose-hispid with simple hairs. Leaves alternate. Flowers pedicellate in terminal scorpioid cymes or panicles of cymes; bracts small, the lowest leafy. Calyx divided almost to the base into 5 ± non-accrescent linear-ovate lobes. Corolla mauve, blue or green, campanulate, broadened above the insertion of the stamens, widened and naked at the throat; lobes 5, usually much longer than the tube, ovate, acuminate, imbricate, spreading and revolute after flowering. Stamens 5, inserted about the middle or near base of tube; filaments short, linear, narrowly ellipsoid or round, sometimes inflated, with a basal triangular or oblong appendage, sometimes reduced or represented by thickened ciliate ridges; anthers linear, with elongate aristate terminal thong-shaped appendages longer than the anther-thecae and coherent along their margins, not twisted. Ovary 4-lobed; style gynobasic, filiform; stigma capitate. Nutlets 1–4, 3-angled or pyramidal, erect, acute, angular, verrucose with basal areole, dorsally keeled. Seed straight; embryo straight with thick ovate plano-convex cotyledons; radicle superior.

A genus of 15 species in NE. Africa including SW. Arabia and Socotra, Zaire, Zambia and Angola; 3 occur in the Flora area, one belonging to subgen. *Cystostemon* and 2 to *Austrovaupelia* A.G. Miller & Riedl.

1. Anther-appendages 1–1.3 cm. long; filaments linear, unexpanded; stems with very closely placed erect narrow leaves; subshrubby herb ± 1.2 m. tall (subgen. *Cystostemon*) . . . .   3. **C. barbatus**
   Anther appendages 3–6 mm. long; filaments narrowly elliptic to round, ± appendaged; stems with less closely placed, less strictly erect and often rather wider leaves; annual or biennial herbs 15–60 cm. tall (subgen. *Austrovaupelia* Miller & Riedl) 2

* 1025 cited by Gibbs and Edmondson is a *Lychnis*. Perry & McNeill (Taxon 36: 483–492 (1987)) have investigated the correct name of this species with involved detail and state that LINN 191.14 is type-material of *E. plantagineum* and its acceptance mandatory over the figure chosen but this is not so.

2. Flowers blue, mauve or reddish; basal filament-appendage
   developed and filament as wide as anther   . . . .   1. **C. hispidus**
   Flowers greenish cream; basal filament-appendage
   inconspicuous and filament much narrower   . . .   2. **C. virescens**

1. **C. hispidus** (*Bak. & Wright*) *A.G. Miller & Riedl* in Notes Roy. Bot. Gard. Edin. 40:14, fig. 4, 3/c, o (1982). Lectotype, chosen by Miller & Riedl: Kenya, Nakuru District, near Lake Elmenteita, *Scott-Elliot* 6640 (K, lecto.!, BM, isolecto.!)

Erect annual or biennial herb, strigose with spreading unpleasantly hispid white bristly hairs, 15–60 cm. tall, with ± woody tap-root; stems much branched, ± woody at the base. Leaves elliptic or oblong, 1.7–7 cm. long, 0.25–1.8 cm. wide, ± acute at the apex, narrowed to the base, densely covered with bulbous-based bristly hairs on both surfaces; petiole 0–3 mm. long. Cymes raceme-like in terminal inflorescences up to 20 cm. long in fruit, densely spreading hispid all over; pedicels 0.8(–2.5) cm. long, elongating to mostly 1.5 cm. in fruit, sometimes red at the apex. Calyx-lobes linear-lanceolate, 5–6(–7) mm. long, densely hispid with bulbous-based hairs on both sides, very slightly enlarging in fruit. Corolla white with pale blue centre, brilliant blue or blue to mauve with red tinge; tube 2–3 mm. long; limb 1.6 cm. wide the lobes ovate-lanceolate, 4.5–7 mm. long, 5 mm. wide, pubescent on midrib outside and inside across the base. Anthers yellow or purple, 3 mm. long, the appendages brown or blackish, 3–4 mm. long, pubescent on back, thecae narrow, widely separated. Style dull mauve, 7–8 mm. long. Nutlets pale brown, obliquely pyramidal, 3.5–4 × 2.5 mm., angled and densely verrucose. Fig. 23.

KENYA. Lake Naivasha, 31 July 1971, *E. Polhill* 135!; Masai District: 3–5 km. SE. along track leaving Nairobi–Magadi road 38 km. from Wilson Airport, 17 May 1981, *Gilbert* 6141! & 6141A! & between Namanga and Kajiado, Nairobi–Namanga road km. 145, 6 Dec. 1961, *Polhill & Paulo* 1008!
TANZANIA. Masai District: E. of Naabi Hill and E. of eastern Serengeti Park boundary, 10 May 1961, *Greenway et al.* 10159! & Engare Nanyuki–Moshi road, 6.4 km. from Arusha–Nairobi road turn-off, 28 Dec. 1961, *Greenway* 10416! & Kitumbeine Mt., 3 Mar. 1969, *Richards* 24269!
DISTR. **K** 1 (see note), 3, 4, 6; **T** 1, 2; not known elsewhere
HAB. Grassland, *Acacia-Commiphora* bushland, and *Tarchonanthus* association on dry volcanic soils; 1050–2100(–2520) m.

SYN. *Trichodesma hispidum* Bak. & Wright in F.T.A. 4(2): 45 (1905)
     Vaupelia hispida (Bak. & Wright) Brand in F.R. 13: 83 (1914); U.K.W.F.: 520 (1974)

NOTE. Newbould (note to his 6008) considers this species to be (together with *Hirpicium diffusum*) a definite indicator of overgrazing and reports that some areas are 'blue with it' for hundreds of metres; he also reports a red dye in the roots. *Lewis* 30 (Northern Frontier Province, Marsabit District, 2°40'N., 36°55'E.) is probably this species but an excessively poor specimen. Further material from this locality is needed.

2. **C. virescens** *A.G. Miller & Riedl* in Notes Roy. Bot. Gard. Edin. 40: 15, fig. 4/3h (1982). Type: Kenya, Northern Frontier Province, 24–27 km. E. of Banessa on Ramu road, *Gillett* 13267 (K, holo.!)

?Annual or biennial or perennial herb 15–40 cm. tall with slender simple but quite woody root; stems ± branched or unbranched, covered with ± dense long spreading white setae (2–)3–4 mm. long. Leaves narrowly oblong-elliptic to lanceolate, 1.5–7.5 cm. long, 0.25–1.4 cm. wide, acute at the apex, gradually narrowed at base into an apparent petiole 0–5 mm. long, with scattered long hairs on upper surface and margins and main nerves beneath, elsewhere very shortly pubescent with minutely bulbous-based hairs. Inflorescence of terminal and axillary cymes ± 2.5 cm. long; bracts leaf-like and bracteoles often exceeding and partly hiding the calyces; pedicels shorter than calyx. Calyx 7–10 mm. long; lobes lanceolate to linear-oblong with scattered bristles and hairs outside, denser and longer inside. Corolla greenish cream; tube 2.5–3 mm. long; lobes ovate, 7–10 mm. long, upwardly curved, narrowing ± abruptly into linear tips 3.5–5.5 mm. long, pilose outside. Stamens 8–10 mm. long; anthers 3.5–5 mm. long with appendages 5–6 mm. long; filament short with expanded portion under ½ the width of the anther; basal filament-appendage inconspicuous, up to 0.5 mm. long. Style 1.25 cm. long. Nutlets not known.

KENYA. Northern Frontier Province: 24–27 km. E. of Banessa on road to Ramu, 22 May 1952, *Gillett* 13267! & 30 km. from Ramu on Malka Mari road, 6 May 1978, *Gilbert & Thulin* 1523!
DISTR. **K** 1; S. Ethiopia

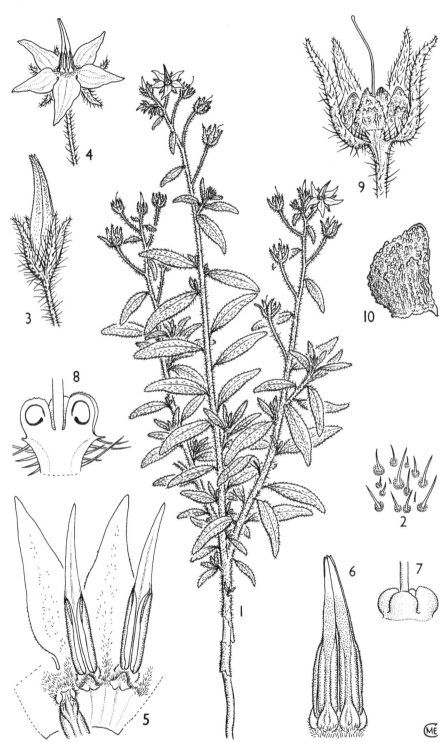

FIG. 23. *CYSTOSTEMON HISPIDUS*—**1**, habit, × ⅔; **2**, portion of upper leaf surface, × 8; **3**, flower bud × 4; **4**, flower, × 2; **5**, stamens, × 6; **6**, portion of corolla and stamens, opened out, × 6; **7**, ovary, × 12; **8**, same, longitudinal section, × 16; **9**, fruit, × 4; **10**, nutlet, showing surface detail, × 8. 1–4, 6–8, from *Polhill* 46; 5, 9, from *Richards* 25645; 10, from *Polhill* 427. Drawn by Mrs Maureen Church.

HAB. *Commiphora-Acacia* and *Commiphora-Sterculia-Terminalia* bushland on thin soil over limestone; 400–810 m.

3. **C. barbatus** (*F. Vaupel*) *A.G. Miller & Riedl* in Notes Roy. Bot. Gard. Edin. 40: 19, fig. 1B (1982). Type: Zaire, Shaba, Mt. Senga, *Kassner* 2925 (B, holo.†, BM, K, iso.!)

Erect subshrubby herb ± 1.2 m. tall; stems ± unbranched, pale brown, densely hairy with spreading white setae and shorter pubescence, very densely leafy. Leaves closely placed, subsessile, erect, linear-lanceolate, 3.5–8 cm. long, 3–6 mm. wide, attenuate to an acute apex, truncate or rounded at the base, the margins strongly revolute, scabrid with long white setae 1–2 mm. long above from rugae bearing rings of cystoliths, similarly densely setose on the midrib beneath but rest of surface with shorter hairs. Panicles terminal and ± narrow, up to 22 cm. long, 9 cm. wide; bracts leaf-like; pedicels (0.3–)0.8–2 cm. long, densely spreading hairy. Sepals narrowly elliptic-lanceolate to lanceolate, 6–9 mm. long, ± 2 mm. wide, densely hairy outside and at margins, very densely long-hairy inside. Corolla pale blue; tube cylindrical, 3–6 mm. long, glabrous outside, with a ring of long hairs above the insertion of the filaments; lobes linear-lanceolate from a triangular base, 1.5–2 cm. long, 3 mm. wide at base, long-acuminate, densely adpressed silky pilose outside but margins ± glabrous. Anthers 4–7 mm. long with appendages 1–1.3 cm. long, 0.7–0.9 mm. wide, acuminate, sparsely pubescent outside. Ovary glabrous, 4-lobed; style 1.5–2 cm. long. Nutlets subtriquetrous, glabrous with outer face ± rounded, ± rugulose and wrinkled.

TANZANIA. Kigoma District: Uvinza, E. of Lugufu, km. 1159 on Central Railway line, 9 Feb. 1926, *Peter* 36627!
DISTR. **T** 4; Zaire, Angola
HAB. ? Woodland; 1060 m.

SYN. *Trichodesma barbatum* F. Vaupel in E.J. 48: 528 (1912); De Wild. in Ann. Mus. Congo, Bot., sér. 4, 2: 128 (1913) & in Ann. Soc. Sci. Brux. 37: 90 (1913) & Contr. Fl. Katanga: 162 (1921)
  Vaupelia barbata (F. Vaupel) Brand in F.R. 13: 83 (1914); Taton, F.C.B., Borag.: 46, t. 6, map 15 (1971)

12. **MYOSOTIS**

L., Sp. Pl.: 131 (1753) & Gen. Pl., ed. 5: 63 (1754)

Annual or perennial erect usually hairy herbs. Leaves alternate, mostly oblong or lanceolate, entire, the basal ones shortly petiolate, the cauline ones sessile. Flowers small in scorpioid cymes. Calyx campanulate, 5-toothed or divided to just below the middle into 5 mostly narrow lobes, persistent, not or slightly accrescent in fruit. Corolla blue, white or yellow; tube cylindric or funnel-shaped, the throat usually with scales, papillae or swellings; lobes 5, spreading, contorted in bud. Stamens 5, included or exserted; filaments filiform; anthers linear, with the connective prolonged into a small trulliform appendage. Disc absent. Ovary deeply 4-lobed; style gynobasic, filiform; stigma obtuse. Fruit composed of 4 erect, oblong-ovoid glabrous shining nutlets fixed by a small basal areole to the flat or convex receptacle.

Cosmopolitan genus with about 50 species mostly in the temperate regions of the Old World, particularly Europe and New Zealand but few in Africa.

Erect plants, loosely tufted but not forming cushions; stems
  elongated and mature inflorescences much longer than
  the leaves:
  Mostly annual with slender erect stems; calyx over 3 mm.
    long exceeding the fruiting pedicel; corolla-limb 1–1.5
    mm. wide . . . .   1. *M. abyssinica*
  Mostly perennial with more robust ascending stems; calyx
    under 2.5 mm. long, shorter than the fruiting pedicel;
    corolla-limb 3–7 mm. wide . . . .   2. *M. vestergrenii*
Densely tufted cushion-forming herb with very short
  procumbent stems and inflorescences hidden amongst
  the leaves . . . .   3. *M. keniensis*

1. **M. abyssinica** *Boiss. & Reut.*, Diagn. Pl. Orient. Nov. 2(11): 122 (1849); Bak. & Wright in F.T.A. 4(2): 58 (1905); F.P.N.A. 2: 134 (1947); A.V.P.: 157 (1957); E.P.A.: 784 (1962); Taton, F.C.B., Borag.: 49, map 16 (1971); Kabuye & Agnew in U.K.W.F.: 523, fig. on p. 522 (1974). Types: Ethiopia, Simien, Demerki, *Schimper* 1146 (G, syn., BM!, P, S, isosyn.) & Mt. Scholoda, *Schimper* 1889 (G, syn., BM!, K!, P, isosyn.)

Annual erect herb 5–12(–40) cm. high, entirely covered with bulbous-based hairs which are often surrounded with cystoliths; stems simple or slightly branched, slender. Leaves oblong to oblanceolate-oblong, 0.25–3.5 cm. long, 0.6–11 mm. wide, obtuse at the apex, truncate at the base; midrib visible beneath. Flowers small, subsessile, in scorpioid cymes with a few leafy bracts in lower part only. Calyx deeply divided into linear-lanceolate to oblong lobes 1.5–2 mm. long, slightly accrescent in fruit, hispid, some of the hairs hooked. Corolla white or blue, glabrous, with slight constriction between the tube and lobes in dry state; tube cylindrical, 1–1.7 mm. long; lobes oblong to obovate, 0.6–0.8 mm. long, 0.4–0.6 mm. wide; swellings minute, papillate. Stamens included. Style 1.5 mm. long; stigma ± clavate. Nutlets dark brown or black, 4 or often reduced to 1, compressed-ovoid, 1.2–1.4 mm. long, 0.8–1 mm. wide, shining, narrowly winged at the apex; areole basal, pale. Fig. 24/1–8, p. 88.

UGANDA. Mbale District: Elgon, W. slope above Budadiri [Butadiri], track to Mudangi through the caldera, 8 Dec. 1967, *Hedberg* 4565! & Elgon, *Benham*!
KENYA. Trans-Nzoia District: Elgon, Kiptogot Waterfall, Sept. 1957, *Tweedie* 1898!; Elgeyo District: N. Cherangani Hills, Chepkotet, Aug. 1968, *Thulin & Tidigs* 236!; Laikipia District: Aberdares Range, National Park, Chebuswa [Eland Hill], 28 Oct. 1970, *Mabberley* 381!
TANZANIA. Moshi District: Kilimanjaro, foot of Kifinika, Nov. 1893, *Volkens* 1341! & SE. side, 20 Mar. 1934, *Schlieben* 4975!; Njombe District: Buanji [Wanji], Magoye [Mwagoje]. 16 Feb. 1914, *Stolz* 2526!
DISTR. U 3; K 3, 4; T 2, 7; Bioko [Fernando Po], Cameroon, Zaire, Rwanda, ?Sudan, Ethiopia

HAB. *Erica-Hypericum-Protea-Alchemilla* bushland, grassland and moorland at lower edge of forest, streamsides; 2000–3250(–3960) m.*

SYN. *M. hispida* Schlechtend. var. *bracteata* A. Rich., Tent. Fl. Abyss. 2: 88 (1850). Type: Ethiopia, Mt. Scholoda, *Schimper* III 1889 (P, holo., BM!, K!, P, iso.)
    M. aequinoctialis Bak. in K.B. 1894: 29 (1894). Type: Tanzania, Kilimanjaro, *Johnston* 123 (K, holo.!, BM, iso.!)

2. **M. vestergrenii** *Stroh* in Beih. Bot. Centralb. 61B: 328 (1941); A.V.P.: 158, 316 (1957); E.P.A.: 784 (1962); Kabuye & Agnew in U.K.W.F.: 523 (1974). Type: Ethiopia, Simien, Demerki, *Schimper* 1152 (P, lecto.**, BM!, K!, S, isolecto.)

Mostly perennial herb 20–60 cm. tall, ± stoloniferous, the stems sometimes flushed crimson. Leaves elliptic to oblong or oblong-oblanceolate, 1–7.5 cm. long, 0.5–1.5 cm. wide, obtuse but apiculate at the apex, rounded to cuneate at base, the cauline leaves smaller, the basal ones strongly narrowed into a winged petiole to 2 cm. long, adpressed pubescent, above often with dense cystoliths at base of hairs, eventually glabrous save for margins and midrib. Flowers in tight densely pubescent head-like cymes ± 1 cm. long, terminal and paired on main and also some lateral branches, eventually becoming up to 10 cm. long; pedicels eventually 6–8 mm. long. Calyx 2–2.2(–3) mm. long; lobes 1.2–1.4(–2) mm. long. Corolla pink in bud then bright blue, whitish yellow in centre; tube ± 1.5 mm. long with papillate gibbosities; lobes ± 2 mm. long, 2.2 mm. wide. Nutlets compressed-ovoid, 1.5 mm. long, 1 mm. wide, shining. Fig. 24/9–17, p. 88.

UGANDA. Mbale District: Elgon, Sasa Camp, 16 Apr. 1950, *Forbes* 259! & Bulambuli, Apr. 1930, *Liebenberg* 1635! & near summit, Jan. 1918, *Dummer* 3357!
KENYA. Trans-Nzoia District: Mt. Elgon crater, 11 May 1948, *J. Bally* 472! & Aberdares, Satima [Sattima], 19 Mar. 1922, *R.E. & T.C.E. Fries* 2602!; N. Kavirondo District: Elgon Caves, 23 Dec. 1956, *Bickford* 46!
TANZANIA. Mbeya District: Kitulo [Elton] Plateau, 24 Jan. 1961, *Richards* 14140!; Rungwe District: Mwakaleli, 8 Jan. 1914, *Stolz* 2411!; Njombe District: Kitulo [Elton] Plateau, Ipume R., 8 Jan. 1957, *Richards* 7592!
DISTR. U 3; K 3, ?4, 5, ?6; T 7; SE. Egypt (fide Cufodontis), Ethiopia, South Africa

---

* Hedberg gives 2700–4000 m. for Kilimanjaro based on Baker's description "13000–14000 ped." but Johnston's original label clearly states 13,200 ft. but may not be accurate.
** Hedberg has designated a lectotype from the 5 specimens of this number at P.

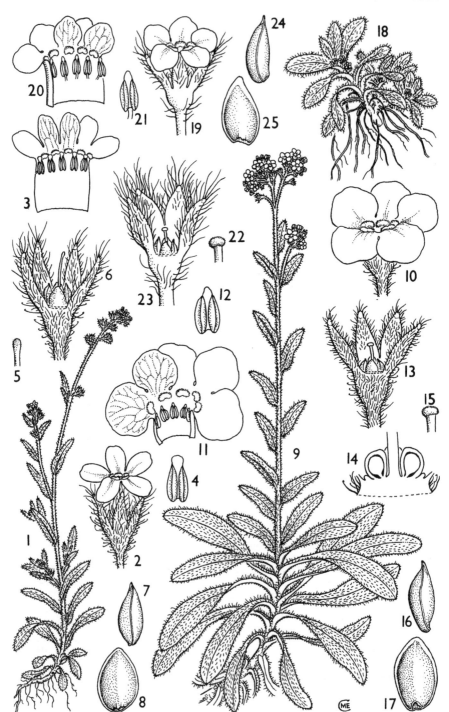

Fig. 24. *MYOSOTIS ABYSSINICA*—**1**, habit, × ⅔; **2**, flower, × 8; **3**, corolla, opened out, × 8; **4**, anther, × 18; **5**, stigma, × 24; **5**, calyx, with young fruit, × 10; **7, 8**, nutlet, two views × 12. *M. VESTERGRENII*—**9**, habit, × ⅔; **10**, flower, × 6; **11**, corolla, opened out, × 6; **12**, anther, × 18; **13**, calyx, with young fruit, × 10; **14**, ovary, longitudinal section, × 24; **15**, stigma, × 24; **16, 17**, two views of nutlet, × 12. *M. KENIENSIS*—**18**, habit, × ⅔; **19**, flower, × 8; **20**, corolla, opened out, × 8; **21**, anther, × 18; **22**, stigma, × 24; **23**, calyx, with young fruit, × 10; **24, 25**, two views of nutlet, × 12. 1, from *Thulin & Tidigs* 236; 2–8, from *Mabberley* 381; 9–12, 14, 15, from *Hedberg* 971; 13, from *Adamson* 472; 16, 17, from *Bogdan* 4467; 18–25, from *Hedberg* 4423. Drawn by Mrs Maureen Church.

HAB. Moist ground along streams, moist moorland, upper *Hagenia-Rapanea* woodland, *Arundinaria* stands, extending into heath zone and the true alpine region, sometimes on rocky ledges; 2000–4250 m.

SYN. [*M. sylvatica* sensu A. Rich., Tent. Fl. Abyss. 2: 89 (1850); Engl. in Hochgebirgsfl. Trop. Afr.: 354 (1892); Bak. & Wright in F.T.A. 4(2): 57 (1905), as "*silvatica*", pro parte, *non* Hoffm.]

NOTE. Differs from the European *M. sylvatica* in pollen morphology (see Hedberg A.V.P.: 316). A closely allied species occurs on Mt. Cameroon; Milne-Redhead (adnot.) is convinced it is distinct.

3. **M. keniensis** *T.C.E. Fries* in N.B.G.B. 8: 420, fig. 7 (1923); A.V.P.: 158 (1957); Kabuye & Agnew in U.K.W.F.: 523 (1974). Type: Kenya, Mt. Kenya, W. slope, 4700 m., *R.E. & T.C.E. Fries* 1399 (UPS, holo., BR, K!, S, iso.)

A low cushion-forming procumbent perennial plant ± 1–5 cm. long and wide, spreading pilose with white hairs, and with creeping rhizome; stems very short or ± absent. Basal leaves oblong, oblong-lanceolate, ovate or spathulate, 0.4–3 cm. long, 0.3–0.9 cm. wide, obtuse, gradually narrowed into a longish petiole to 1.3 cm. long; upper leaves rather smaller, oblong, very shortly petiolate, all long-ciliate at margins and adpressed pilose beneath. Flowers few, almost hidden in the leaf-axils, subsessile or with very short spreading hairy pedicels. Calyx 2.5 mm. long becoming 3 mm., divided to the middle; tube campanulate, ± glabrous, lobes lanceolate, 1.7 mm. long, 1 mm. wide, with densely ciliate margins, adpressed pilose near apex outside but ± glabrous at base and inside. Corolla-tube yellow or cream, equalling the calyx, 1.2 mm. long; limb pale blue or purple-blue, ± 1.6 mm. diameter with rounded lobes ± 1 mm. long and 0.7 mm. wide. Anthers 0.7 mm. long including a minute apical appendage. Fruiting pedicels up to 1 mm. long. Nutlets ellipsoid, 1.5 mm. long, 0.7 mm. wide, acute, with keeled margin, shining. Seeds under 1 mm. long. Fig. 24/18–25.

KENYA. Mt. Kenya, Upper Hausburg [Hausberg] Valley, 12 Jan. 1965, *Allt* 7! & Upper Hinde Valley, 4 Feb. 1965, *Allt* 61! & W. slope, along the Naro Moru track between Kampi ya Farasi and Two Tarn Col, 24 Nov. 1967, *Hedberg* 4423!
DISTR. **K** 4; Ethiopia (Bale, Mt. Batu)
HAB. Gregarious in bare soil, solifluction terraces, under boulders, with grass tussocks in *Dendrosenecio* associations, etc.; 4200–4700 m.

13. ECHIOCHILON

Desf., Fl. Atlant. 1: 166, t. 47 (1798); I.M. Johnston in Journ. Arn. Arb. 38: 261 (1957) (detailed revision)

Leurocline S. Moore in J.B. 39: 257, t. 424 (1901)

Tetraedrocarpus Schwartz in Mitt. Inst. Bot. Hamburg 10: 212 (1939)

Usually perennial (one species a herbaceous annual) with spreading or ascending branched stems, becoming woody with age, sparsely to densely hairy, the leaf-hairs often arising from well-developed mineralized discoid bases. Leaves all cauline, mostly alternate but lowest on a shoot opposite and in one species all opposite, rather thick or somewhat succulent, narrow and elongate, mostly without visible nervation, the margins somewhat involute. Inflorescences few–many-flowered, bracteate, with flowers interspersed among leaves along the outer half of the leafy shoot or aggregated into erect elongating pseudo-racemose clusters terminating the leafy stems and branches, never distinctly scorpioid; pedicels short and stout or obsolete. Calyx-lobes 5, oblong to lanceolate or oblanceolate, equal or very unequal, sometimes united at base into a very short tube. Corolla white, lilac, blue or reddish purple, sometimes pale but becoming blue or purple with age, salver-shaped or funnel-shaped, symmetrical to strongly zygomorphic, outside glabrous to strigulose or pubescent or with minute stipitate glands, hairy at the throat; lobes rounded, obovate, ovate or semicircular, equal or unequal. Stamens 5 with unequal filaments, included or shortly exserted. Style short or extending only to the throat, never exserted, slender, entire or in one species very shortly bilobed at the apex. Nutlets 1–4, grey to reddish, broadly lanceolate, ovate or cordate in dorsal outline, dorsally not or imperfectly ridged, ventrally obtuse or with median keel.

A genus of 17 species mostly of arid areas extending from S. Kenya to Ethiopia, Somalia, N. Africa, Arabia, Sinai, Palestine, Baluchistan and S. Iran.

FIG. 25. *ECHIOCHILON LITHOSPERMOIDES*—**1**, habit, × ⅔; **2**, portion of leaf, lower surface, × 4; **3**, flower, × 3; **4**, corolla, opened out, × 4; **5**, ovary, × 16; **6**, ovary, longitudinal section, × 20; **7**, stigma, × 20, **8**, from *Bogdan* 5498; **9**, nutlet, showing surface detail, × 8. 1–7, from *Napper* 1723; 8, from *Bogdan* 5498; 9, from *Kokwaro* 2899. Drawn by Mrs Maureen Church.

E. lithospermoides (*S. Moore*) *I.M. Johnston* in Contr. Gray Herb. 73: 50 (1924) & in Journ. Arn. Arb. 38: 273 (1957); E.P.A.: 787 (1962); Kabuye & Agnew in U.K.W.F.: 523 (1974). Types: Kenya, Laikipia, *Gregory* (BM, syn.) & Ethiopia, Gof & between Lè and Tocha, *Delamere* (BM, syn.!)

Shrublet, quite woody at the base, forming patches 10–60(–100) cm. long and wide, the stems much branched, spiky and tufted, the plant reaching a height of 10–40 cm., ± glaucous. Leaves lanceolate, 1–3 cm. long, 2–5 mm. wide, gradually narrowed to acute or acuminate apex, ± narrowed at base, usually glabrous above, pustulate on lower surface or with short spreading bulbous-based hairs, ± costate, margin usually coarsely hispid-ciliate, almost appearing toothed. Inflorescence an erect unilateral cyme, elongating with age, 10–30 cm. long, 10–50-flowered, with numerous bracts the lowest resembling the leaves, very gradually reduced upwards; pedicels 1–2 mm. long or obsolete. Calyx-lobes unequal, lanceolate, 4–8 mm. long, 0.4–1.3 mm. wide, ± twice as large at maturity, hispid-ciliate, inner surface glabrous, outer pustulate or with bulbous-based hairs. Corolla rose, blue turning pink or purple with red spots, strongly zygomorphic, 1.2–1.5 cm. long, tubular in lower half, expanding above into an extremely oblique funnel-like throat and limb 7–9 mm. wide, glabrous outside, with a zone of hairs inside 4–6.5 mm. above base; lobes rounded, unequal, up to 2.5 mm. wide. Stamens borne at unequal heights; filaments 0.2–0.4 mm. long; style 3–4 mm. long; stigmas oblique, surmounted by 2 laterally compressed sterile style-tips. Nutlets reddish, 2–2.4 mm. long and wide, dorsally tuberculate, ventrally angular with triangular areole ± 1 mm. wide contracted into an upwardly directed sulcus. Fig. 25.

KENYA. Northern Frontier Province: Banissa, 27 June 1951, *Kirrika* 92!; Laikipia District: Rumuruti, 12 Aug. 1952, *Bogdan* 3519!; Masai District: Amboseli National Reserve, on Amboseli–Emali road, 10 Feb. 1964, *Napper* 1723!
DISTR. **K** 1, 3, 6; Ethiopia
HAB. Sparse *Acacia* woodland and bushland, grassland with or without scattered trees, also recorded from temporarily flooded mud flats, both on red and black soils; 850–1850 m.

SYN. *Leurocline lithospermoides* S. Moore in J.B. 39: 257, t. 424/1 (1901)
Lobostemon lithospermoides (S. Moore) Bak. & Wright in F.T.A. 4(2): 60 (1905)

14. **TRICHODESMA**

R.Br., Prodr. Fl. Nov. Holl.: 496 (1810); Brand in E.P. IV. 252 (78): 19–44, figs. 2–4 (1921); Brummitt in K.B. 37: 429–450 (1982) & in K.B. 40: 851–854 (1985), *nom. conserv.*

Herbs or subshrubs, hispid, hairy or sometimes almost glabrous, erect. Leaves alternate or subopposite. Flowers pedicellate in lax sometimes many-flowered terminal cymose inflorescences. Calyx deeply divided into 5–6 imbricate ovate to lanceolate lobes which are often winged, narrowed to cuspidate at the apex, rounded to cordate at the base, accrescent. Corolla blue or white, subrotate or funnel-shaped; tube with naked throat; lobes 5–6, ovate, ± cuspidate at the apex. Stamens 5–6, with very short and flattened filaments; anthers oblong or linear-lanceolate, ± pubescent on back, nearly always forming a cone with the connectives produced above the thecae in a linear usually twisted appendage. Ovary ovoid, 4-lobed. Style gynobasic or terminal, long and subulate; stigma small, subglobose. Fruits with 4 ovoid nutlets, 3-angled on the ventral face or one discoid sometimes margined nutlet, smooth, rugose or hairy on the back. Seeds subglobose or obovoid.

About 45 species in tropics and subtropics of Africa, Asia and Australia.

Brand divided the genus into 6 sections and some of the rather striking differences could be considered almost of generic importance but Brummitt & Bidgood (personal communication) could find no supporting palynological differences. Four sections occur in East Africa.

Sect. **Trichodesma**
Leiocarya Hochst. in Flora 27: 29 (1844)
Trichodesma R.Br. sect. *Leiocaryum* A.DC. in DC., Prodr. 10: 172 (1846); Gürke in E. & P. Pf. IV. 3a: 99 (1893), pro parte; Brand in E.P. IV. 252 (78): 38 (1921)
Fruit with 4 emarginate smooth shining nutlets (species 1–3)

Sect. **Acanthocaryum** *Brand* in E.P. IV. 252 (78): 27 (1921)
   Fruit with 4 nutlets with distinct serrate margins and other faces with glochidiate setae (species 4, 5)

Sect. **Serraticaryum** *Verdc.* sect. nov. ab sectionibus alteris nuculis fructus 4 ovoideo-cupuliformibus margine late explanatis serratisque intus basi fundo glochidio-setosis differt. Typus: *Trichodesma trichodesmoides* (Bunge) Gürke
Trichodesma sect. *Friedrichsthalia* Brand in E.P. IV. 252 (78): 28 (1921), *non Friedrichsthalia* (Fenzl) A. DC., *nom. illegit.*
   Fruit with 4 nutlets which are ovoid-cup-shaped with distinct serrate margins and glochidiate hairs on the outer face, i.e. base of cup (species 6).
   Brand in E.P. IV. 252 (78): 34 (1921) puts *T. calathiforme*, one of the synonyms of 6, *T. trichodesmoides* (see page 99), in sect. *Ommatocaryum* A.DC. in DC., Prodr. 10: 174 (1846), but the type of that, *T. aucheri* A.DC., has very different nutlets; Brand puts other synonyms of *T. trichodesmoides* in sect. *Friedrichsthalia* in his sense.

Sect. **Friedrichsthalia** (*Fenzl*) *A.DC.* in DC., Prodr. 10: 173 (1846); Gürke in E. & P. Pf. IV. 3a: 99 (1894), pro parte; Brummitt in K.B. 37: 437 (1982), *non* Brand in E.P. IV. 252 (78): 28 (1921)
Friedrichsthalia Fenzl in Endl. & Fenzl, Nov. Stirp. Dec.: 53 (1839)
Trichodesma R.Br. sect. *Trichocaryum* Brand in E.P. IV. 252 (78): 21 (1921), *nom. illegit.*
   Fruit with a single depressed circular cushion-shaped silky hairy nutlet (species 7, 8).

Fruit with 4 ± erect glabrous or setose nutlets:
   Nutlets smooth and shiny; margins not raised nor spinous
      (sect. *Trichodesma*):
      Calyx-lobes ± truncate at the base . . . .   1. *T. zeylanicum*
      Calyx-lobes hastate, sagittate or at least cordate at the base:
         Anthers joined into a clearly visible exserted cone with
            connective-appendages twisted together; leaves
            with fine pubescence and longer hairs on the
            venation . . . .   2. *T. indicum*
         Anthers separate and hidden in the tube, the connective-
            appendages straight, apparently never twisted
            together; leaves with only longer hairs on the
            venation . . . .   3. *T. inaequale*
   Nutlets with winged serrate margins, the teeth drawn out and
      faces with glochidiate setae (sect. *Acanthocaryum*):
      Anthers in a distinct well-exserted shortly hairy cone;
         indumentum of stem, pedicel, and sometimes leaf,
         spreading . . . .   4. *T. hildebrandtii*
      Anthers ± enclosed in corolla-tube, not in a distinct
         exserted cone or if apparently so then obscured with
         long hairs; indumentum of stem, pedicel and leaf
         adpressed:
         Leaves 2–6.5 × 0.25–1.6 cm.; fruiting sepals very
            distinctly cordate at the base . . . .   5. *T. marsabiticum*
         Leaves 2–10.5 × 0.8–6.5 cm.; fruiting sepals not distinctly
            cordate at the base (sect. *Serraticaryum*) . . .   6. *T. trichodesmoides*
Fruit of a single depressed silky hairy cushion-shaped nutlet
   (sect. *Friedrichsthalia*):
   Stems glabrous or with few setae on lower internodes;
      pedicels ± glabrous; sepals (3–)3.5–6(–7) mm. broad,
      lanceolate to ovate, rounded to cordate at the base in
      fruit; corolla white or pink . . . .   7. *T. physaloides*
   Stems usually setose-pubescent; pedicels with mostly
      spreading or hooked hairs, sometimes with only
      adpressed hairs; sepals 5–8(–9) mm. wide in flower, ±
      distinctly cordate in fruit; corolla blue, mostly pinkish in
      bud . . . .   8. *T. ambacense*

   1. **T. zeylanicum** (*Burm. f.*) *R. Br.*, Prodr. Fl. Nov. Holl.: 496 (1810); A.DC., Prodr. 10: 172 (1846); A. Rich., Tent. Fl. Abyss. 2: 91 (1850); Bot. Mag. 80, t. 4820 (1854); Bak. & Wright in F.T.A. 4(2): 51 (1905); Z.A.E. 2: 280 (1911); Brand in E.P. IV. 252 (78): 40, fig. 4/D–E (1921);

FIG. 26. *TRICHODESMA ZEYLANICUM*—1, habit, × ⅔; 2, part of upper leaf surface, × 4; 3, flower, × 3; 4, corolla, opened out, × 4; 5, ovary and part of calyx, × 4; 6, ovary, longitudinal section, × 14; 7, stamen, × 6; 8, fruit, × 1; 9, nutlet, × 4; 10, seed, × 6. 1–7, from *Milne-Redhead & Taylor* 9481; 8, 9, from *Milne-Redhead & Taylor* 7204; 10, from *Verdcourt* 2364. Drawn by Mrs Maureen Church.

F.P.S. 3: 99 (1956); E.P.A.: 780 (1962); Ivens, E. Afr. Weeds: 83, fig. 40 (1967); Taton, F.C.B., Borag.: 37, map 12 (1971); Kabuye & Agnew in U.K.W.F.: 520 (1974); Brummitt in F.Z. 7(4): 95 (1990). Type: Ceylon, *Garcin* in *Herb. Burm.* (G, lecto.!)

Perennial, less often annual, erect and usually branched herb or subshrubby herb 0.3–1.5(–2.1) m. tall from a taproot; stem ± woody at the base, densely pubescent and scabrid with tubercle-based hairs 1–2.5 mm. long. Upper leaves sessile, alternate, oblong to oblong-lanceolate, 2–12.5(–18) cm. long, 0.6–3.2(–5.5) cm. wide, narrowed to the apex, rounded to subcordate at the base, discolorous, scabrid with tubercle-based hairs above, the tubercles ringed with cystolith-cells, rough beneath, densely pubescent to grey-tomentose with tubercle-based hairs on the nerves; lower leaves opposite or subopposite, similar but larger, up to 16 cm. long, 5 cm. wide with a petiole 0.2–1 cm. long. Flowers drooping, in lax terminal many-flowered cymes; bracts similar to the leaves, 1.2–3.5 cm. long, 0.3–1 cm. wide, often cordate at the base; pedicels reddish, slender, 0.5–3.5 cm. long, with both short and long hairs. Calyx-lobes ovate-lanceolate, 7–10 mm. long, 3.2–3.5 mm. wide, enlarging in fruit to 1.6–2 cm. long, acute to acuminate at the apex, pubescent inside, grey-tomentose and with intermingled long tubercle-based hairs particularly on the midnerve and margins outside. Corolla pale to deep blue with white or pink to purple centre, with a dark reddish purple spot at the base of each lobe, or lobes white along mid-area and with blue outer areas and with reddish purple area below where the two blue areas join; tube funnel-shaped, 4–5 mm. long, glabrous outside, inside at the base on both sides of anther-insertions with cushions of scales; lobes 5–6, broadly ovate, 4–4.5 mm. long, 3.8–6 mm. wide, with an abrupt narrowly triangular acute twisted acumen 2.5–3 mm. long. Stamens sessile; anthers lanceolate, thecae 2.5 mm. long, with a tuft of hairs at the base and connective-appendages as long as the thecae or up to 4 mm., twisted together. Ovary 1.5–2 mm. diameter, glabrous, 4-lobed; style 6–8 mm. long, with obscurely subglobose stigma. Nutlets 4, grey marbled brown (? sometimes black), compressed ovoid, 4 mm. long, 3.2 mm. wide, the external face slightly convex, shining; internal face 3-angled, rugose and tuberculate. Figs. 26; 27/1, p. 96; 28/1, p. 98.

UGANDA. W. Nile District: 1.6 km. E. of Omugo Rest Camp, 10 Aug. 1953, *Chancellor* 143!; Karamoja District: Mt. Kadam [Debasien], Amaler, Jan. 1936, *Eggeling* 2525!; Teso District: Soroti, near Kapiri Ferry, 16 Sept. 1954, *Lind* 376!

KENYA. Meru District: 16 km. SE. of Meru, Giaki Experimental Farm, 16 Aug. 1958, *Bogdan* 4630!; Nyanza Basin, *Battiscombe* 494!; Teita District: Tsavo National Park East, Galana R., E. of Lugard's Falls, about 53 km. from Voi Gate, 1 Jan. 1967, *Greenway & Kanuri* 12924!

TANZANIA. Tanga District: Gereza [Gereze] E., 30 Aug. 1964, *Semsei* 3874!; Rufiji District: Mafia I., Kanga–Kikuni, 14 Aug. 1937, *Greenway* 5104!; Songea District: shore of Lake Malawi [Nyasa], Mbamba Bay, 5 Apr. 1956, *Milne-Redhead & Taylor* 9481!; Pemba I., Vitongoge, 21 Sept. 1929, *Vaughan* 668!

DISTR. **U** 1–3; **K** 4–7; **T** 1–8; **Z**; **P**; Sudan, Ethiopia, Mozambique, Malawi, Zambia, Zimbabwe, Angola and South Africa (Transvaal, Natal), also Comoros, Madagascar, Mascarenes, India, Ceylon, Malaya, Java, Philippines and Australia

HAB. Grassland, *Commiphora*, *Grewia*, etc., bushland, a common weed in old and new cultivations and a pioneer on disturbed ground, on both well-drained red loam and marshy ground on black cotton soil; 7–1710 m.

SYN. *Borago zeylanica* Burm. f., Fl. Indica: 41, t. 14/2 (1768); L., Mant. Pl. Alt.: 202 (1771); Jacq., Ic. Pl. Rar. 2, t. 314 (1789)
Leiocarya kotschyana Hochst. in Flora 27: 30 (1844). Types: Ethiopia, Djelajeranne, *Schimper* 625 (B, syn. †, BM, K, isosyn.!) & Modat, Aguar, *Schimper* 1025 (B, syn.†, K, isosyn.!) & Sudan, Nubia, Camamil and Gebbel Kassan [Fazokl on sheet], *Kotschy* 542 (B, syn.†, K, W, isosyn.!)
Boraginella zeylanica (Burm. f.) O. Kuntze, Rev. Gen. Pl. 2: 435 (1891)
Borraginoides zeylanica (Burm. f.) Hiern, Cat. Afr. Pl. Welw. 1: 720 (1898)

NOTE. Brand (loc. cit.: 41) divides this species into 2 subspecies, one confined to Australia; the typical subspecies he divides into 2 varieties, one Australian and one African and the typical variety into two forms forma *typicum* (i.e. forma *zeylanicum*) and forma *longifolium* Brand (loc. cit.: 42), based on *Engler* 976 from the W. Usambaras, *Stuhlmann* 76 from Dar es Salaam, *Stuhlmann* 237 from Mpwapwa and *Meebold* 8817 from India. I have not considered forma *longifolium* worth recognising and since it is otherwise the typical form which occurs in East Africa have not investigated the Australian variants.

2. **T. indicum** (*L.*) *Lehm.*, Pl. Fam. Asperif.: 193 (1818); A.DC. in DC., Prodr. 10: 172 (1846); Boiss., Fl. Orient. 4: 280 (1875); Bak., Fl. Maurit. & Seych.: 203 (1877); C.B. Cl. in Fl. Brit. Ind. 4: 153 (1883); Sedgwick in Rec. Bot. Surv. India 6: 347 (1919); Brand in E.P. IV. 252(78): 38 (1921) (full synonymy); Verdc. in K.B. 44: 701 (1989). Type: specimen grown in Uppsala Garden, *Linnaeus* (LINN 188/2, lecto.!)

Bushy much-branched spreading or ± erect herb 14–35 cm. tall from a woody rootstock; young stems sometimes reddish, with short pubescence and white bristly tubercle-based hairs, older with peeling epidermis. Leaves elliptic or oblong, 2.5–5.5 cm. long, 0.3–1.5 cm. wide, rounded to subacute at the apex, amplexicaul at the base, sessile, with bristly hairs with ring of cystoliths around their base and often densely grey-pubescent. Flowers drooping in lax few–many-flowered cymes or solitary; pedicels recurved, 0.8–1.2 cm. long, spreading hairy; bracts similar to the leaves but much smaller. Calyx-lobes triangular, 0.8–1.5 cm. long, 4–8 mm. wide, hastate at the base, with coarse tubercle-based white bristly hairs on margins and midnerve and pubescence as well, not enlarging. Corolla pale blue inside, sky blue outside, centre dark red, with yellowish brown oblong mark at the base of each lobe; tube funnel-shaped, 4–5 mm. long; lobes ovate, 4–7 mm. long, 3–6 mm. wide, abruptly and narrowly cuspidate; joined part of the limb ± 3 mm. long, with brown patches of different texture below each sinus. Anthers whitish, the cone 8 mm. long, exserted ± 6 mm. including the twisted glabrous connective-appendage tips; cone shortly densely hairy outside; base of anthers with hair tufts, otherwise glabrous. Style 6.5–8.5 mm. long. Nutlets splitting off and leaving a quadrangular central column, 4, grey-buff or with dark brown spots, compressed ovoid, 5 mm. long, 3 mm. wide, smooth and shining outside, strongly corrugated inside. Figs. 27/2, p. 96; 28/2, p. 98.

KENYA. Mombasa, 24 Apr. 1927, *Linder* 2650! & 14 Oct. 1955, *Milne-Redhead & Taylor* 7101! & Freretown, 30 May 1934, *Napier* 3311 in *C.M.* 6226!
TANZANIA. Uzaramo District: Dar es Salaam, link road by University, 10 m. from road N. of approach road to Ubungo transmitter, 2 Oct. 1976, *Wingfield* 3687! & Dar es Salaam, Selander Bridge, 22 Jan. 1971, *Harris* 5570!
DISTR. **K** 7; **T** 6; Mauritius, Réunion, Afghanistan, Pakistan, India, Philippines; presumably introduced into East Africa
HAB. Grassland on sand, sand above high-water mark, waste ground, coconut plantations, roofs of old houses; 0–45 m.

SYN. *Borago indica* L., Sp. Pl.: 137 (1753)
   Trichodesma amplexicaule Roth, Nov. Pl. Sp. Ind. Or.: 104 (1821). Type: India, *Heyne* (B, holo.†)

3. **T. inaequale** *Edgew.* in Journ. Asiat. Soc. Bengal 21: 175 (1852); Verdc. in K.B. 44: 700 (1989). Type: India, Banda, *Edgeworth* (holo. ubi?, K, iso.!?*)

More or less erect annual herb 10–60 cm. tall, the stems densely covered with spreading bristly hairs and shorter pubescence as well. Leaves narrowly oblong-elliptic or oblong-lanceolate to slightly oblanceolate, 2.5–15 cm. long, 1.2–2.5 cm. wide, obtuse or subacute at the apex, rounded truncate or subcordate at the base, sometimes amplexicaul, densely covered with tubercle-based hairs above arising from rings of cystoliths and on the venation beneath but without additional pubescence. Flowers in few-flowered cymes and solitary flowers in axils of uppermost 3 leaves (foliaceous bracts); pedicels 1.3–2.5(–4) cm. long, with long and short pubescence. Calyx-lobes cordate to hastate at the base, 0.9–1.3 cm. long, 0.5–1.3 cm. wide, acuminate, setose and pubescent, with bristly margins and midrib. Corolla blue or ?pink or red; tube cylindrical, 5–5.5 mm. long; limb funnel-shaped, 4 mm. long including the broad often slightly unequal lobes 2–2.5 mm. long, 2.5–3 mm. wide. Stamens quite separate, the anthers not cohering in a cone, 2.8–4 mm. long including 1.2 mm. long connective-appendages which are not twisted; connective-hairs reaching almost to the top of the appendages or only slightly beneath. Style 3.5–4.5 mm. long. Nutlets 4, buff with darker grey markings, compressed-ovoid, 5 mm. long, 3.5 mm. wide, 2.8 mm. thick, the outer surface smooth, the inner with raised rugae which are closer and more elevated at apex of nut. Figs. 27/3, p. 96; 28/3, p. 98.

TANZANIA. Singida District: Itigi–Singida road, 19.2 km. from Singida, 28 Mar. 1965, *Richards* 20011!
DISTR. **T** 5; Cameroon, Sudan, Ethiopia, India
HAB. Grassland on sand; 1500 m.

SYN. [*T. amplexicaule* sensu A.DC. in DC., Prodr. 10: 172 (1846), pro parte**; C.B. Cl. in Fl. Brit. Ind. 4: 154 (1883), pro parte; Sedgwick in Rec. Bot. Surv. India 6: 347 (1916); Brand in E.P. IV. 252 (78): 39 (1921), pro parte; Gamble in Fl. Madras 2: 899 (1923), pro parte, *non* Roth]
   T. uniflorum Brand in F.R. 12: 504 (1913) & in E.P. IV. 252 (78): 39 (1921); Heine in F.W.T.A., ed. 2, 2: 323 (1963). Types: Cameroon Mt., 'Johann-Albrechts-Hütte', *Winkler* 1255 (B, syn.†) & Limbe [Victoria], *Zahn* 530 (B, syn.†)

* Specimen is labelled "*Trich. amplexicaulis* 6027 Banda, Edgeworth".
** A.DC. cites Roth but his description refers to this present species.

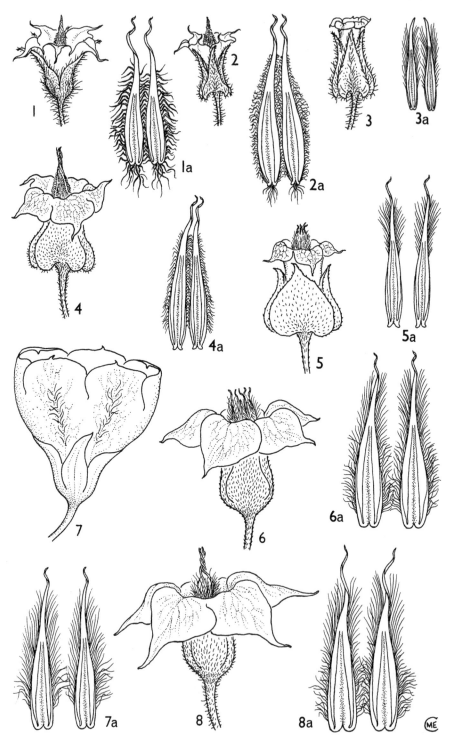

FIG. 27. *TRICHODESMA SPP.*—flower × 2; **a**, stamens × 6 of **1**, *T. ZEYLANICUM*; **2**, *T. INDICUM*; **3**, *T. INAEQUALE*; **4**, *T. HILDEBRANDTII*; **5**, *T. MARSABITICUM*; **6**, *T. TRICHODESMOIDES*; **7**, *T. PHYSALOIDES*; **8**, *T. AMBACENSE* subsp. *HOCKII*. 1, from *Milne-Redhead & Taylor* 9481; 2, from *Milne-Redhead & Taylor* 7101; 3, from *Richards* 20011; 4, from *Mathew* 6650; 5, from *Carter & Stannard* 661; 6, from *T. Adamson* K13; 7, from *Tweedie* 1991; 8, from *Richards* 6914. Drawn by Mrs Maureen Church.

[*T. indicum* sensu Hutch. & Dalz., F.W.T.A. 2: 200 (1931), adnot. and some Indian authors, *non* (L.) Lehm.]

T. sedgwickianum Banerjee in Bull. Bot. Soc. Bengal 16: 10 (1963); Banerjee & Pramanik in Bull. Bot. Surv. India 17: 116, figs. A–D on p. 117 (1978), *nom. invalid.** Type not stated

NOTE.   Edgeworth's description fits this plant well but I have not yet managed to trace the holotype in any Indian herbarium.

4. **T. hildebrandtii** *Gürke* in E. & P. Pf. IV. 3a: 99, t. 40/G, H (1894) *nomen* & in N.B.G.B. 1: 61 (1895); Bak. & Wright in F.T.A. 4(2): 48 (1905); Brand in E.P. IV. 252 (78): 27 (1921); E.P.A.: 779 (1962); Brummitt in K.B. 40: 853 (1985). Type: Somalia (N.), Ahl Mts., *Hildebrandt* 847a (B, holo.†)

Woody based herb or subshrub 25–50 cm. tall; stems with spreading tubercle-based hairs which wear down to just white tubercles plus fine deflexed pubescence; epidermis pale brown or whitish. Leaf-blades ovate or oblong-ovate, 1.7–7(–9) cm. long, 1–2.7(–4) cm. wide, ± acute at the apex, truncate then narrowed or cuneate at the base, scabrid with tubercle-based hairs above and less so beneath mainly on the nerves; petiole short, 3–10 mm. long. Flowers pendent in 2–5(–10)-flowered inflorescences; pedicels 1–1.6 cm. long with upwardly directed or spreading white bristly hairs and shorter hairs. Calyx-lobes ovate-lanceolate, 1–1.2 cm. long, 4–6.5 mm. wide, becoming 1.5–2 × 1–1.7 cm., often joined for ± 8 mm., acuminate, deeply cordate at the base, 1-nerved, with short stiff adpressed hairs and curved ones on margins. Corolla white with purple-brown centre margined yellowish; tube cylindrical-funnel-shaped, 5–8 mm. long; limb reflexed, ± 7.5 mm. long overall, the lobes broadly rhombic-ovate, narrowed at base, 5 mm. long including 3 mm. long twisted acumen, 6 mm. wide, with 2 brown elliptic areas 2 × 1.2 mm. (dry state) below each sinus. Anthers joined in a well-exserted 8 mm. long cone, 1 cm. long including (2–)4–5 mm. long twisted glabrous connective-appendages; tops of anthers outside and lower parts of appendages shortly hairy; adnate part of filaments below anther-bases shortly pubescent; anther-thecae 5 mm. long, glabrous. Ovary 1.2 mm. tall, 2 mm. wide, ± 4-lobed; style 5.5–12 mm. long. Nutlets ovate, 6–7 mm. long, dorso-ventrally compressed, margined, marginally very distinctly spinous-dentate, dorsally with rigid spreading spines which are retrorsely shortly pubescent, ventrally very shortly grey-pubescent and venose. Figs. 27/4; 28/4, p. 98.

KENYA.   Northern Frontier Province: Malka Mari, Daua R., 27 June 1951, *Kirrika* 97! & S. Turkana, near Mugurr, 6 June 1970, *Mathew* 6650!; Turkana District: hill N. of Lodwar, 18 May 1954, *Popov* 1567!

DISTR.   **K** 1, 2; Ethiopia, Somalia and Arabia (Oman)

HAB.   *Acacia–Commiphora* bushland, steep limestone valleys by watercourses, stony plains, sometimes in crevices of basement rocks; 600–850 m.

SYN.   *T. cardiosepalum* Oliv. in Hook., Ic. Pl., t. 2436 (1896); Brand in E.P. IV. 252 (78): 27 (1921). Type: Arabia, Oman, foot of the Dhofar Mts., *Bent* 115 (K, holo.!)

5. **T. marsabiticum** Brummitt in K.B. 40: 851 (1985). Type: Kenya, Northern Frontier Province, Marsabit, Gof Choba, E. rim, *Bally & Smith* 14856 (K, holo.!, BR, EA!, iso.)

Shrub 0.3–1.6 m. tall; young stems with sparse short adpressed tubercle-based setae, the older with peeling grey epidermis revealing chestnut bark beneath. Leaf-blades narrowly oblong-elliptic to elliptic-lanceolate, 2–6.5 cm. long, 0.25–1.6 cm. wide, bluntly acute at the apex, the actual tip very narrowly rounded, narrowly cuneate at the base, covered with short adpressed hairs from circular base of cystoliths; petiole 0.2–1 cm. long. Inflorescences 2–8-flowered, 1–3 cm. long; bracts robust or lowest flowers in axils of reduced leaves; pedicels ± 1–2(–3 in fruit) cm. long, ± glabrous or sparsely adpressed setose. Sepals broadly ovate, up to 1.4 cm. long, 9.5 mm. wide, shortly acuminate at the apex, rounded to slightly cordate at the base, enlarging in fruit to 2.2 cm. diameter, ± round and distinctly cordate, shortly adpressed setose when young and with puberulous margins. Corolla white with orange-brown marks below the sinuses; tube 1.2–1.3 cm. long, broadest at the middle, narrowed at base, densely hairy inside near insertions of the

---

* Banerjee states nom. nov. and cites *T. amplexicaule* A.DC. but A.DC. does cite Roth and has merely misidentified the material he had. Banerjee is thus effectively describing a new species; he gives a Latin description but cites no type.

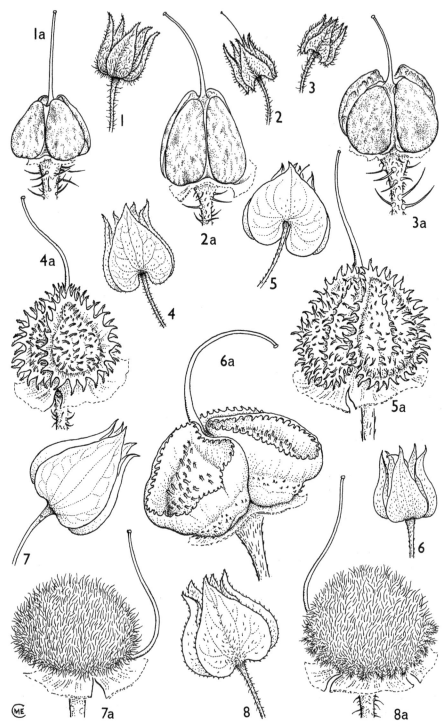

FIG. 28. *TRICHODESMA SPP.*—fruit × 1; **a**, nutlet, × 4 of **1**, *T. ZEYLANICUM*; **2**, *T. INDICUM*; **3**, *T. INAEQUALE*; **4**, *T. HILDEBRANDTII*; **5**, *T. MARSABITICUM*; **6**, *T. TRICHODESMOIDES*; **7**, *T. PHYSALOIDES*; **8**, *T. AMBACENSE* subsp. *HOCKII*. 1, from *Milne-Redhead & Taylor* 7204; 2, from *Gillett* 20312; 3, from *Richards* 20011; 4, from *Gilbert & Thulin* 1439; 4a, from *Bone* 410; 5, from *Oteke* 5; 5a, from *Bally & Smith* 14856; 6, from *T. Adamson* K13; 7, from *Eggeling* 2587; 7a, from *Tanner* 5104a; 8, from *Semsei* 62; 8a, from *Richards* 6442. Drawn by Mrs Maureen Church

filaments, the lobes broadly triangular, 2–5 mm. long, 3–6 mm. wide, reflexed or not, acuminate. Anthers not in a distinct exserted cone, 6 mm. long, narrowed above into linear connective-appendages 5 mm. long; anther glabrous save for apical part and lower part of connective-appendages both of which are densely long-pilose; tips of appendages glabrous for 1–2 mm., at first twisted. Ovary 1 mm. tall, 3 mm. wide, ± 4-lobed; style subulate, 8 mm. long, tapering above. Nutlets 4, compressed ovate, ± 6 mm. long 4.5 mm. wide and 1.5 mm. thick, glochidiate, the marginal glochidia broader and ± joined together to form a serrated crest; inside smooth save for a marginal ring of short setae. Figs. 27/5, p. 96; 28/5.

KENYA. Northern Frontier Province: 11 km. N. of Marsabit, S. of Choba, 24 Nov. 1977, *Carter & Stannard* 661! & Mt. Marsabit, Gof Redo, 2 June 1960, *Oteke* 5!; Meru District: North Meru Grazing Scheme, Mula Hills, Mar. 1960, *Katumani Exp. Farm* 49!
DISTR. **K** 1, 4; not known elsewhere, but see note
HAB. Steep lava slopes amongst volcanic boulders; 1075–1230 m.

NOTE. Closely allied to the southern African *T. angustifolium* Harv. *Linson & Gieson* 86 (Meru District, Isiolo, Lewa Downs Farm, 0°20'N, 37°45'E, 1685 m.) has very narrow leaves 2–3.2 mm. wide and narrowly acuminate sepals, thus reducing the difference between *T. marsabiticum* and *T. angustifolium* but possible floral differences in corolla-lobes and anther-appendages need assessment with better material of the Kenyan taxon. *Gillett & Beckett* 23312 from central Somalia is very similar to *T. marsabiticum* and perhaps only infraspecifically distinct, some of the leaves being similar to those of *T. marsabiticum* but others narrowly ovate up to 2.8 cm. wide.

6. **T. trichodesmoides** (*Bunge*) *Gürke* in E. & P. Pf. IV. 3a: 99 (1894); E.P.A.: 780 (1962); Brummitt in K.B. 40: 853 (1985). Type: Ethiopia, without exact locality, possibly Adua, *Schimper* 710 (?LE, holo., BM!. E!, K!, iso., ?P, iso.)

Herb ± 60 cm. tall; stem and all parts (save corolla) ± densely, less often sparsely, covered with adpressed upwardly directed stiff white hairs from circles of cystoliths; stems also with finer pubescence. Leaf-blades elliptic to ovate, 2–10.5 cm. long, 0.8–6.5 cm. wide, ± acute at the apex, cuneate at the base; petiole 0.6–3.5 cm. long. Inflorescence ± 8–10-flowered; pedicels ± 1.7 cm. long. Calyx-lobes narrowly ovate-lanceolate, 0.8–1.3 cm. long, 4–4.5 mm. wide, becoming 1.8–2.3 × 1.1–1.6 cm. in fruit, acuminate, joined at base, scarcely narrowed, with upwardly directed slightly tubercle-based white hairs. Corolla white to pale lilac or cerise to mauve with greenish to brownish centre or a dark ring around mouth of tube with brown markings between the lobes; tube ± 1 cm. long; lobes triangular, 6 mm. long, 8 mm. wide, shortly or sometimes long-acuminate. Anthers densely hairy, not in a distinct exserted cone, the tips exserted 4–5 mm.; connective-appendages filiform, glabrous, sinuous, at first twisted together. Style 1.4 cm. long. Nutlets 4, ovoid-bath-shaped, 5–7 mm. long, 3–3.5 mm. wide, 3–4 mm. thick, hollow, the margin inrolled and conspicuously toothed, the teeth often tipped with a tuft of hairs; inside with a median row or covered all over with stout glochidiate setae or sometimes quite smooth; the base of the free wall-like part often has roughening outside. Figs. 27/6, p. 96; 28/6, p. 98.

KENYA. Northern Frontier Province; 12 km. SW. of Marsabit, near Karsadera, 24 Nov. 1977, *Carter & Stannard* 635! & Marsabit, July 1958, *T. Adamson* 13! & 14 km. S. of Loiyangalani, 3 June 1980, *Kuchar* 13391!
DISTR. **K** 1; Ethiopia, Somalia, Djibouti and SW. Arabia
HAB. Stony hillsides, tufa slopes with low scrub; 370–1200 m.

SYN. *Friedrichsthalia trichodesmoides* Bunge, Del. Sem. Hort. Dorpat. 1843: 7 (1843) & in Linnaea 18: 152 (1844)
Trichodesma calathiforme Hochst. in Flora 27: 29 (1844); Bak. & Wright in F.T.A. 4(2): 49 (1905); Brand in E.P. IV. 252 (78): 34 (1921). Type as for *T. trichodesmoides**
Streblanthera trichodesmoides (Bunge) A. Rich., Tent. Fl. Abyss. 2: 92 (1850)
?*S. oleifolia* A. Rich., Tent. Fl. Abyss 2: 93, t. 78 (1850). Type: Ethiopia, Chelicut, *Petit* (P, holo., K, iso.!) (as '*oleaefolia*')
Boraginella trichodesmoides (Bunge) O. Kuntze, Rev. Gen. Pl. 2: 436 (1891)
Trichodesma grandifolium Bak. in K.B. 1894: 29 (1894); Bak. & Wright in F.T.A. 4(2): 49 (1905); Brand in E.P. IV. 252 (78): 33 (1921). Type: Somalia, Adda Galla and Zafarag (Lafarug), *James & Thrupp* s.n. (K, holo.!)
T. pauciflorum Bak. in K.B. 1894: 29 (1894); Bak. & Wright in F.T.A. 4(2): 48 (1905); Brand in E.P. IV. 252 (78): 43 (1921). Type: Ethiopia, Red Sea, Hanish [Harnish] I., *Slade* (K, holo.!)
T. schimperi Bak. in F.T.A. 4(2): 47 (1905). Type: Ethiopia, Ebenat, *Schimper* 1472 (K, lecto.!, BM, E, isolecto.!)

* Hochstetter cites *Schimper* 710 (B†) but also specimens from Selassaquila 1838 and Djelajeranne (3 Nov. 1839)

?*T. oleifolium* (A. Rich.) Bak. in F.T.A. 4(2): 47 (1905), as '*oleaefolium*'; Brand in E.P. IV. 252 (78): 34 (1921)

T. calathiforme Hochst. var. *schimperi* (Bak.) Brand in E.P. IV. 252 (78): 35 (1921)

T. trichodesmoides (Bunge) Gürke var. *schimperi* (Bak.) Cuf., E.P.A.: 780 (1962), *nom. invalid.*

NOTE. I think *T. oleifolium* is merely a form of *T. trichodesmoides* but only the type has been seen. A sterile sheet, *J. Adamson* in *E.A.H.* 10893, said to be a rhododendron-like shrub, from South Island in Lake Turkana [Rudolf] has short stout internodes with peeling epidermis but is I feel sure only an extreme form of this species caused by the exceptionally exposed saline conditions. *Linder* 3619 (20 km. S. of Loiyangalani, 100 m. above Lake Turkana, 28 Apr. 1986) seems to be a form of this species with green shoots from grazed stems; the internodes are short and thick with peeling papery epidermis and the leaves small as in the type of *T. pauciflorum*.

7. **T. physaloides** (*Fenzl*) *A.DC.* in DC., Prodr. 10: 173 (1846); Gürke in E. & P. Pf. IV. 3a: 99, t. 40/F (1894); Wright in Fl. Cap. 4(2): 11 (1904); Bak. & Wright in F.T.A. 4(2): 46 (1905), pro parte; Brand in E.P. IV. 252 (78): 22 (1921); Brenan in Mem. N.Y. Bot. Gard. 9: 6 (1954); F.P.S. 3: 88 (1956); Taton, F.C.B., Borag.: 39, t. 5, map 13 (1971); Kabuye & Agnew in U.K.W.F.: 521, fig. (1974); Brummitt in K.B. 37: 439, fig. 1, map 1 (1982) & in F.Z. 7(4): 96, t. 28 (1990). Type: Sudan, Fazokl, *Kotschy* 577 (BM, K, iso.!)

Perennial herb with 1–several erect stems, some flowering and some separate and sterile, from a fleshy or woody tuberous rootstock up to 30 cm. long and 5 cm. diameter; stems ± glabrous or with sparse setae below. Leaves sessile, rather thickish, variable, broadly ovate or elliptic to lanceolate or linear-lanceolate to linear-elliptic, (1–)2–6(–9) cm. long, (0.2–)0.5–2.5(–3.5) cm. wide (or rarely, on sterile stems, up to 11 × 4.2 cm.), ± acute at the apex, cuneate to rounded at the base, the lower ± opposite and upper ± alternate, with ± dense circles of cystoliths above often bearing minute setae above but with fewer beneath or lacking save on the main nervation. Inflorescence 2–30 cm. long and up to 30 cm. wide, with 2–9(–12) primary branches each 1–8(–12)-flowered; pedicels 1–2.5(–3.2) cm. long, elongating to 4.5 cm. in fruit, glabrous or with a few tubercles just below the flower; bracts lanceolate to linear, 3–7 mm. long. Flowers faintly scented. Sepals lanceolate to ± ovate, (0.8–)1–1.6(–1.8) cm. long, (3–)3.5–6(–7) mm. wide, enlarging to 2.7 × 2.2 cm. in fruit, becoming rounded or cordate at the base, glabrous or with tubercles at the base and round the margin and sometimes sparsely on the surfaces also. Corolla-tube white with brown marks at sinuses between the lobes which are white or sometimes cream or tinged pink; tube funnel-shaped, 1.1–1.3 cm. long, hairy inside at level of anther-insertion; lobes broadly ovate, 5–8 mm. long, 1–1.4 cm. wide with 1.5–2 mm. long apicule, glabrous or more usually densely pilose down the middle outside. Anthers oblong-lanceolate, 1–1.4 cm. long, villous on back so that the connectives are surrounded by long flexuous hairs for about ½ their length, the appendages shorter than or almost as long as the thecae. Ovary 1.5–2 mm. diameter, with 4 distinct lobes; style filiform, glabrous with minute stigma. Nutlet discoid, 0.9–1.4 cm. diameter, 3 mm. tall, densely hairy with ± dentate margins. Seed brown, discoid, 7–8 mm. diameter, 1.5 mm. thick, with pattern of openly reticulate but not raised veins. Figs. 27/7, p. 96; 28/7, p. 98.

UGANDA. Karamoja District: Kadam [Debasien], near Amaler, Jan. 1936, *Eggeling* 2587!; Teso District: Serere, Mar. 1932, *Chandler* 579!; Mbale District: Elgon, Siroko valley, 15 Feb. 1924, *Snowden* 821!
KENYA. W. Suk District: Kanyem–Kacheliba road, 7 May 1934, *Mortimer* 275!; Uasin Gishu, Apr. 1932, *Harvey* 155!; N. Kavirondo District: SW. slopes of Elgon, Mar. 1934, *Tweedie* 138!
TANZANIA. Arusha District: track from Arusha National Park to Sakila, 22 Mar. 1971, *Richards & Arasululu* 26812!; Mpanda District: 16 km. N. of Kasogi, 6 Aug. 1959, *Harley* 9170!; Mbeya District: Mbozi, 28 Aug. 1933, *Greenway* 3613!
DISTR. U 1, 3; **K** 2, 3, 5; **T** 1–8; S. Sudan, W. Ethiopia, Burundi, S. Zaire, Zambia, Malawi, Mozambique, Zimbabwe, Swaziland and South Africa (Transvaal and Natal)
HAB. *Brachystegia* woodland, *Protea-Combretum*, etc., wooded grassland, bushland, grassland and sometimes in cultivations, particularly on burnt patches after rains; 1050–2400 m.

SYN. *Friedrichsthalia physaloides* Fenzl in Endl. & Fenzl, Nov. Stirp. Dec.: 54 (1839)
Boraginella physaloides (Fenzl) O. Kuntze, Rev. Gen. Pl. 2: 435 (1891), as "*physalodes*"
Borraginoides physaloides (Fenzl) Hiern, Cat. Afr. Pl. Welw. 1: 721 (1898)
Trichodesma droogmansianum De Wild. & Th. Dur. in B.S.B.B. 39, 4: 69 (1900); Bak. & Wright in F.T.A. 4(2): 47 (1905); Brand in E.P. IV. 252 (78): 23 (1921). Type: Zaire, Lualaba, *Descamps* (BR, holo.)
T. glabrescens Gürke in E.J. 30: 389 (1901); Bak. & Wright in F.T.A. 4(2): 48 (1905). Type: Tanzania, Njombe District, Ukinga Mts., Ussangu, *Goetze* 1267 (B, holo.†, BR, E!, iso.)
T. 'frutescens' K. Schum. in Just's, Bot. Jahresb. 29: 564 (1902), error for *T. glabrescens*

T. ringoetii De Wild. in F.R. 13: 100 (1914); Brand in E.P. IV. 252 (78): 22 (1921). Type: Zaire,
Shaba, Nieuwdorp, *Ringoet* 6 (BR, holo.!)
T. droogmansianum De Wild. & Th. Dur. var. *glabrescens* (Gürke) Brand in E.P. IV. 252 (78): 24
(1921)

8. **T. ambacense** *Welw.* in Apont. Phytogeo.: 589 (1859) as '*ambacensis*'; Brummitt in K.B.
37: 442 (1982) & in F.Z. 7(4): 98 (1990). Type: Angola, *Welwitsch* 5450 (LISC, holo., BM, BR,
COI, K!, iso.)

Perennial herb with 1–several erect stems to 50(–70) cm. tall from a woody rootstock;
stems ± setose-pubescent with ± irregularly spreading soft setae and often softer smaller
hairs interspersed or occasionally ± glabrous with sparse stiffer tubercular setae or very
rarely ± entirely glabrous. Leaves sessile or subsessile, broadly ovate to elliptic or
oblanceolate or linear-oblong, (2–)4–7(–9) cm. long, (0.3–)0.7–3(–3.6) cm. wide, or
sometimes on sterile shoots up to 16 × 5 cm., acute or obtuse at the apex, cuneate to
rounded at the base, the lower opposite, the upper usually subopposite, alternate, with flat
circles of cystoliths on both surfaces with or without hairs or with short seta-like hairs
without tubercle bases, often ± densely hairy beneath when young. Inflorescences 2–30
cm. long, with 2–9(–12) primary branches each bearing 1–8(–12) flowers; pedicels 1–3(–
3.5) cm. long, becoming 6(–8) cm. in fruit, with spreading hairs, often hooked at the tips,
or occasionally with only sparse to ± dense upwardly directed hairs; bracts narrowly ovate,
0.4–1.3 cm. long, 1–4 mm. wide. Sepals narrowly lanceolate to ovate, (0.8–)1.1–1.7(–2) cm.
long, (2.5–)3–8(–9) mm. wide, enlarging to 2.8 cm. broad in fruit and then cordate at the
base, sparsely to densely hairy, with or without conspicuous tubercles, closely or loosely
upwardly adpressed in the upper part but often spreading or reflexed towards the base,
rarely subglabrous with a line of tubercle hairs around the margin. Corolla-tube white, or
cream and lobes white suffused with pale grey-blue to a spectacular gentian blue with
greenish brown or brown marks at the base of each lobe at sinuses; tube 0.7–1.1 cm. long,
inside with pads of scaly hairs on each side of anthers; lobes broadly ovate-triangular to
broadly ovate, 0.75–1.5 cm. long, 0.5–1.2(–1.5) cm. wide, spreading or reflexed, cuspidate
or abruptly apiculate, in herbarium specimens usually obscuring the sepals in a majority
of fully opened forms, usually glabrous but rarely with a line of hairs down the middle of
the lobes. Anthers 1.2–1.6 cm. long, hairy on the back, the connective-appendages almost
as long as the thecae, surrounded by long hairs or not. Ovary clearly divided into 4 lobes.
Nutlet ± 1 cm. in diameter, silky hairy, with ± dentate margin.

SYN. *Boraginella ambacensis* (Welw.) O. Kuntze, Rev. Gen. Pl. 2: 435 (1891)
     Trichodesma angolense Brand in E.P. IV. 252 (78): 26 (1921); Taton, F.C.B., Borag.: 44 (1971), *nom.*
     illegit. Type as for *T. ambacense*

subsp. **hockii** (*De Wild.*) *Brummitt* in K.B. 37: 446 (1982) & in F.Z. 7(4): 98 (1990). Type: Zaire, Shaba,
Lubumbashi [Elisabethville], *Hock* (BR, holo., K, photo.!)

Leaves broadest at or below their middle, often rounded at the base, sessile. Pedicels 1–2.2(–2.7)
cm. long, elongating to 3.5(–4.5) cm. in fruit. Sepals ovate, (1–)1.2–1.6(–1.8) cm. long, (3–)5–8(–9) mm.
wide, usually broadly overlapping each other and the margins often curved outwards, enlarging to
2.8 cm. wide in fruit. Staminal hairs extending beyond the fertile part so that the connectives are
surrounded by long flexuous hairs for half their length or more. Figs. 27/8, p. 96; 28/8, p. 98.

UGANDA. Acholi District: Aswa, Gulu, Feb. 1943, *E. Forbes* 52!
KENYA. Aberdares, Sept. 1951, *Sharpe* in *Bally* 8835! (see note)
TANZANIA. Ufipa District: Lake Kwela, 9 Nov. 1950, *Bullock* 3484!; Chunya District: SE. of Lake
     Rukwa, Mlupa-Sira [Lupa], 14 Oct. 1932, *Geilinger* 3044! & 3037!; Iringa District: Itaka, 31 Aug. 1935,
     Greenway 3652!
DISTR. U 1; K 3; T 4, 7; Nigeria, Cameroon, Sudan, SE. Zaire, Zambia, Malawi, N. Mozambique,
     Zimbabwe and Botswana
HAB. *Brachystegia* and other secondary woodland, burnt grassland particularly at forest edges;
     780–2100(–3000?) m.

SYN. *T. ledermannii* F. Vaupel in E.J. 48: 529 (1912); Brand in E.P. IV. 252 (78): 24 (1921); Heine in
     F.W.T.A., ed. 2, 2: 323 (1963). Type: Cameroon, near Laro, *Ledermann* 3080 (B, holo.†);
     Cameroon, Ourosangé, *Lowe* 3350 (K, neo.!)
     T. hockii De Wild. in F.R. 11: 546 (1913); Brand in E.P. IV. 252 (78): 26 (1921); Brenan in Mem.
     N.Y. Bot. Gard. 9: 6 (1954); Taton, F.C.B., Borag.: 44 (1971)
     T. tinctorium Brand in B.J.B.B. 4: 393 (1914) & in E.P. IV. 252 (78): 23 (1921). Type: Zaire, Shaba,
     Lukafu, *Verdick* 104 (BR, holo.)
     T. verdickii Brand in B.J.B.B. 4: 392 (1914) & in E.P. IV. 252 (78): 23 (1921). Type: Zaire, Shaba,
     Lukafu, *Verdick* 140 (BR, holo.)

NOTE. The Kenya specimen cited above is said to be from 10000 ft. but this is surely an error, in fact the locality seems highly unlikely and should be viewed with suspicion unless confirmed. Many atypical specimens have been seen. Some are intermediates between the subspecies in having very narrow sepals, e.g. *Sanane* 227 (Mpanda District, road to Uruwira, 12 July 1968 at 1050 m.) and appear very different. *Richards & Arasululu* 25999 (from about the same place at 974 m.) and several other specimens from **T** 4 are similar. Apart from these, intermediates between *T. physaloides* and *T. ambacense* subsp. *hockii* occur in areas where the distributions overlap, e.g. *Mahinde* 211 (Mpanda District, Utahya, 4 Sept. 1958), *Jefford & Newbould* 1201 (Mpanda District, Mahali Mts., Lumbye, 30 July 1958) and *Geilinger* 3067 (Chunya District, Sinipale, 15 Oct. 1952). It is tempting to regard these as hybrids but there is no real evidence. They have more abundant hairs on the stem but are otherwise good *T. physaloides*. *Azima* 688 (Kigoma District, 8 km. S. of Ilagara, 23 Aug. 1963) is similar.

15. CYNOGLOSSUM

L., Sp. Pl.: 134 (1753) & Gen. Pl., ed. 5: 65 (1754); Brand in E.P. IV. 252 (78): 114 (1921)

Perennial, biennial or rarely annual usually very hairy herbs or subshrubby herbs. Leaves alternate, the radical ones long-petiolate, the cauline ones sessile or shortly petiolate. Cymes terminal or axillary, generally ebracteate, scorpioid, lengthening considerably in fruit. Flowers white or blue, pedicellate or subsessile; calyx persistent with 5 spreading or reflexed lobes, slightly accrescent. Corolla cylindric-rotate or funnel-shaped with 5 broadly ovate imbricate spreading obtuse lobes; throat closed by ± squarish or crescent-shaped appendages. Stamens 5, included, inserted at the base or towards the apex of the tube. Ovary distinctly 4-lobed, adhering to the central stylar column by a part of the internal face only of each lobe; style gynobasic; disc sometimes distinct and lobulate. Nutlets 4, depressed-ovoid, usually densely echinulate with glochidiate spines or smooth in part, sometimes the apical beak cohering with the style.

An almost cosmopolitan genus of about 50 species of which 8 occur in the Flora area.

Popov has suggested the genus should be split and many tropical and subtropical species he refers to *Paracynoglossum* Popov (Fl. U.R.S.S. 19: 717 (1953)). This is not accepted by Riedl (Fl. Turkey 6: 306 (1978)) and does indeed seem unnecessary. Mill & Miller (Notes Roy. Bot. Gard. Edin. 41: 473–482 (1984)) dealing with Arabian species do accept *Paracynoglossum*. I prefer to treat it as a subgenus. The coherence of the beak of the nutlet to the style seems an inadequate character and, moreover, varies somewhat in some African species.

The genus *Cynoglossum* is well known to be one of the most intractable; in Africa and India it is particularly so and is undoubtedly the most difficult genus of the family in the Flora area. Apart from the distinctive species the main core has usually been called *C. coeruleum* or *C. geometricum* according to the fruit ornamentation which, however, proves to be very variable; other promising characters such as corolla-size, sepal-shape, style-length and leaf-shape are all variable but certainly define some populations. Single populations I have examined have always been very uniform in most characters. The epithet *coeruleum* was erroneously claimed to be a later homonym by Mill & Miller (in Notes Roy. Bot. Gard. Edin. 41: 474 (1984)). The African assemblage seems to be the result of interaction between European elements related to *C. montanum* L. and Indian elements related to *C. zeylanicum* (Lehm.) Brand* and *C. wallichii* G.Don. Populations similar to one or both or even nearly identical can be found in N. Ethiopia and E. Tanzania but it has been thought unwise to introduce names from outside Africa without a great deal more work — only the utmost confusion would result. *C. lanceolatum*, long recognised to occur throughout Africa and Asia, but frequently misidentified, probably complicates matters by hybridisation; this has been suspected independently by B.L. Burtt and myself but no firm proof is available. The often suggested idea that *C. geometricum* is merely a variety of *C. lanceolatum* is quite erroneous — they differ very widely in fruit. I think improvements in the classification will only be possible from a synthesis of detailed local studies in Africa and India and careful comparisons between taxa from both areas and not from general revisions by authors unfamiliar with the actual phytogeography of most of the areas.

C. amabile Stapf & Drummond has been cultivated in East Africa, e.g. Uganda, Masaka City Centre, 23 Apr. 1972, *Lye* 6721! (at 1240 m.). This species also occurs as an escape in NE. Tanzania and is treated fully below (see sp. 1)

1. Leaves ± tomentose with non-scabrid indumentum,
   densely adpressed softly velvety on both surfaces, the
   hairs very short . . . .   1. *C. amabile*

---

* See Verdc. in K.B. 43: 343 (1988) for an elucidation of the correct interpretation of this name.

Leaves with distinctly scabrid indumentum particularly above, the hairs tubercle-based and often with rings of white cystoliths; the hairs eventually often breaking off leaving a rough tuberculate surface or if leaves thinly velvety and ± non-scabrid then hairs coarse, ± 1 mm. long 2

2. Radical and lower stem-leaves wider, elliptic-ovate to ovate, 10–27 × 3.5–17 cm., rounded, cuneate or subcordate at the base; nutlets larger, 5–6 mm. wide . . . . . 2. *C. amplifolium*

Radical and lower stem-leaves narrower, oblong, elliptic, oblanceolate, etc., but never more than narrowly ovate, up to 5.5 cm. wide, always cuneate at the base; nutlets smaller, 2–5.5 mm. wide and if 4–5.5 mm. then leaves only 0.5–3.5 cm. wide 3

3. Flowers usually white, sometimes pale blue or with a blue centre, small, the corolla-tube ± 1 mm. long and lobes 0.5–2 × 0.5–1 mm.; nutlets subglobose, 1.5–2 mm. diameter, the whole fruit only 4–5 mm. wide . . . 3. *C. lanceolatum*

Flowers mostly bright blue (but albinos occur), larger, the tube 1.5–3 mm. long and lobes up to 3 × 3 mm.; nutlets more ovoid and/or much larger, 2–5 mm. diameter, the whole fruit 5–10 mm. wide 4

4. Sepals eventually lanceolate, up to 4–5 × 0.9–1.7 mm. and lowest flowers of cymes with pedicels up to ± 3 cm. long; nutlets with sparse ± scattered glochidia on face . . . . 7. *C. ukaguruense*

Without above characters 5

5. Nutlets ± uniformly covered with glochidia 6

Nutlets with marginal glochidia and along a median line, the other areas on the upper face smooth or with a few tubercles or scattered glochidia 7

6. Calyx-tube distinctly obconic with lobes broadly rounded ovate, 1.5 × 1.2–1.8 mm.; inflorescences ± corymb-like cymes; pedicels 0.2–1.7 cm. long; nutlets 4–5.5 mm. wide, completely uniformly covered with glochidia under 0.5 mm. long . . . . 5. *C. aequinoctiale*

Calyx-tube more ovoid, with lobes more oblong or if ovate mostly larger and more acute; inflorescences less corymb-like; pedicels mostly short, only occasionally up to 1.5 cm. particularly of the oldest central flower of a dichasium; nutlets 3.5–4 mm. wide with usually more slender longer glochidia . . . . 4. *C. coeruleum*

7. Leaves ± thinly velvety with rather long white hairs, eventually ± tubercle-based above; nutlets shallowly bowl-shaped, round to triangular in outline, with a ± uniseriate row of marginal flat narrowly triangular glochidia and a few on the median line, the underside with only small tubercles (U1) . . . . 9. *C. karamojense*

Leaves in no way velvety, the indumentum always scabrid above; nutlets not shaped as above 8

8. Calyx-lobes up to 4–5 × 1.5 mm. after corollas fall; immature fruits with glochidia around the margin and across the median line, with very few on intervening faces; herb to about 90 cm., with corolla-limb up to 8 mm. across (**T2**, Mt. Hanang) . . . . 6. *C. hanangense*

Calyx-lobes usually much shorter after corollas fall or if not then flowers smaller 9

9. Tall perennial herb to 1.8 m. with oblanceolate radical leaves to 30 × 3.5 cm.; flowers in long-pedunculate well-branched compound dichasia of scorpioid cymes which in the early stages are ± capitate; peduncles up to 16 cm. long; corolla-limb up to 1 cm. wide; nutlets greyish glaucous, 2–2.2 × 1.8–2 mm., with short

triangular glochidia on the back and around the
margin and 2–3 across the median ridge, also a few
shorter ones in between (mostly Kenya, Cherangani
Hills)   . . . .   8. *C. cheranganiense*
Mostly shorter annual or perennial herbs to ± 1 m., with
smaller radical leaves and inflorescences more shortly
pedunculate; nutlets mostly wider with more slender
glochidia (widespread)   . . . .   4. *C. coeruleum* var.
                                                    mannii and intermediates

1. **C. amabile** *Stapf & Drummond* in K.B. 1906: 202 (1906); Brand in E.P. IV. 252 (78): 135
(1921); Meikle, Garden Flowers: 310, fig. 97 (1963); Ariza Espinar in Kurtziana 17: 145, fig.
1 (1984). Type: China, Yunnan, Mengtsze, *Hancock* 133 (K, lecto.!)

Annual or biennial herb 0.3–1.2 m. tall, branched from the base; rootstock ± simple,
6–10 cm. long; stems woody at the base, densely adpressed velvety hairy, scarcely scabrid,
± ridged in dry state. Cauline leaves numerous and ± closely placed, narrowly oblong to
lanceolate, 2–20 cm. long, 0.4–4 cm. wide, acute at the apex, sessile or ± amplexicaul at the
base, densely adpressed softly velvety hairy on both surfaces; nervation impressed above,
raised beneath, grey-green when dry; basal leaves up to 9 cm. long, 1.5 cm. wide with
petioles to 4 cm. long. Inflorescences terminal and axillary, forming a compound
inflorescence ± 10 cm. long in young state, extending to 35 cm.; cymes 2-branched, each
branch eventually 5–10 cm. long; pedicels ± 2 mm. long, becoming longer and deflexed in
fruit, up to 4 mm. long. Calyx ± 3 mm. long, densely pubescent; lobes oblong-elliptic to
ovate, 2 mm. long, just over 1 mm. wide, becoming more ovate and somewhat accrescent
and 3 × 2 mm. Corolla intense blue, the limb ± 1 cm. wide; tube 2.5 mm. long, the throat
with retuse finely papillose bosses; lobes rounded, ± 4 mm. diameter. Anthers pale blue, 1
mm. long, shortly protruding between the bosses. Style 2 mm. long, narrowed at the apex.
Nutlets whitish, depressed-ovoid, 3 mm. long, 2–2.5 mm. wide, covered all over with short
glochidia, thickened towards the base, the outer ± forming a marginal serrated crest. Fig.
29/1.

KENYA. Nairobi, Chiromo Estate, 11 Dec. 1970 (cult.), *Mathenge* 732!
TANZANIA.   Lushoto District: Magamba, 23 Jan. 1950, *Faulkner* 511! & same area, near Lake Mvomoi,
   23 Apr. 1968, *Renvoize & Abdallah* 1693! & Sunga, Shagai saw-mill, 17 May 1953, *Drummond &*
   Hemsley 2598!
DISTR.   T3; China (Tibet, Szechwan and Yunnan); cultivated and naturalised in many parts of the
   world
HAB.   Margins of cultivations, forest clearings, grassland on hillsides, etc.; clearly an escape from
   gardens but now naturalised; 1350–1900 m.
NOTE.   Jex-Blake (Gard. E. Afr., ed. 4: 326 (1957)) states it is annual, at least as used in horticulture.
   There is no doubt that this is very closely allied to *C. furcatum* Wall.*, which has smaller flowers,
   and that *C. amabile* var. *parviflorum* Turrill (Bot. Mag. 166, t. 82 (1949)), described from material
   cultivated in Ceylon, is intermediate.

2. **C. amplifolium** *A.DC.* in DC., Prodr. 10: 149 (1846); A. Rich., Tent. Fl. Abyss. 2: 91
(1850); Engl., Hochgebirgsfl. Trop. Afr.: 353 (1892); Bak. & Wright in F.T.A. 4 (2): 53
(1905); Z.A.E. 2: 281 (1911); R.E. Fries, Wiss. Ergebn. Schwed. Rhod.-Kongo-Exped. 1: 272
(1916); Brand in E.P. IV. 252 (78): 141 (1921); F.P.N.A. 2: 132, fig. 6 (1947); E.P.A.: 781
(1962); Heine in F.W.T.A., ed. 2, 2: 324 (1963); Taton, F.C.B., Borag.: 54, map 19 (1971);
Kabuye & Agnew in U.K.W.F.: 521 (1974); Martins in F.Z. 7(4): 103 (1990). Type: Ethiopia,
between Endschetcap and Schoata, *Schimper* 564 (G, holo., BM, K, UPS, iso.!)

Perennial herb or subshrub 0.3–1.8 m. tall, with thick tuberous root to 45 cm. long;
stems annual, erect, at first densely velvety adpressed hairy but sometimes becoming
almost glabrous. Radical and lower stem-leaves elliptic-ovate to ovate, 10–27 cm. long,
3.5–17 cm. wide, acuminate at the apex, rounded, cuneate or subcordate at the base,
short-to long-decurrent into the 9–42 cm. long petiole, sparsely hairy on both faces;
stem-leaves sessile or decurrent and shortly petiolate, elliptic-lanceolate to ovate, 2–18
cm. long, 1.6–10 cm. wide, acuminate at the apex, rounded at the base, above ± subscabrid

* This has incorrectly been sunk into *C. zeylanicum* (Lehm.) Brand; see Verdc. in K.B. 43: 343
(1988).

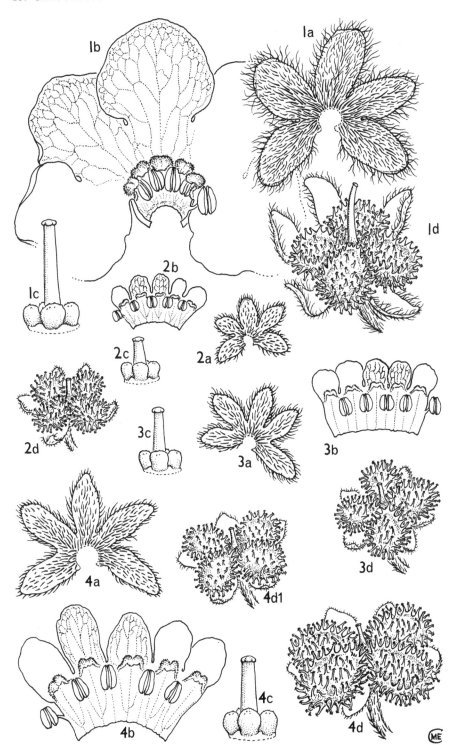

FIG. 29. *CYNOGLOSSUM SPP.*—**a**, calyx × 8; **b**, corolla, × 8; **c**, ovary and style, × 20; **d**, fruit, × 5 of **1**, *C. AMABILE*; **2**, *C. LANCEOLATUM*; **3**, *C. LANCEOLATUM* ? × *COERULEUM*; **4**, *C. COERULEUM* subsp. *JOHNSTONII* var. *JOHNSTONII*. 1a–c, from *Renvoize & Abdallah* 1693; 1d, from *Lye* 6721; 2, from *Battiscombe* 472; 3, from *Hazel* 31; 4a–d, from *Johnston* s.n.; 4d₁, from *Bidgood & Congdon* 138a. Drawn by Mrs Maureen Church.

to less often distinctly scabrid with tubercle-based hairs surrounded by cystoliths, ± velvety hairy beneath, becoming almost glabrous or with sparse hairs on the raised and reticulate venation; petiole 0–2 cm. long. Inflorescences terminal and axillary, the cymes very short, ± 1 cm. long at first, rapidly elongating to 6–12 cm. in fruit; pedicels 2–5(–10) mm. long, lengthening and reflexing in fruit to 1–2.5(–3.5) cm.; all axes adpressed pubescent to velvety, sometimes yellowish in dry state. Sepals broadly elliptic to ovate, 1.5–2.5(–4) mm. long, 1.5–1.7(–3.5) mm. wide, ± acute, accrescent to 5 × 4.5 mm. in fruit, pubescent and ciliate outside, glabrous inside. Corolla light to dark blue, sometimes reddish purple at first and persistently at throat; tube 1.3–2.6 mm. long; lobes broadly elliptic to round or oblate, 1.3–2.5(–4) mm. long, 1.8–3.2(–5) mm. wide; throat-bosses ± square, papillate; filaments 0.2–0.3 mm. long; anthers 0.7–1 mm. long. Ovary ± 1 mm. long; style 0.5–1.2 mm. long; stigma terminal, globose, bilobed. Fruit 0.8–1.5 cm. wide; nutlets cushion-shaped, depressed, 5–6 mm. wide, the external ± convex face uniformly covered all over with short stout conic glochidia or only around the margins and along a median line.

var. **amplifolium**

Nutlets with external face ± uniformly covered with short glochidia. Fig. 30/1–10.

UGANDA. Kigezi District: Mt. Mgahinga, June 1949, *Purseglove* 2940! & Muhavura-Mgahinga saddle, Sept. 1946, *Purseglove* 2139!; Mbale District: Bugishu, Butandiga, 2 Sept. 1932, *A.S. Thomas* 474!
KENYA. Trans-Nzoia District: NE. Elgon, Aug. 1971, *Tweedie* 4097!; Kiambu District: Limuru, 27 June 1909, *Scheffler* 237! (in absence of fruit); Masai District: 32 km. from Narok, W. Mau Forest, 2 Oct. 1954, *Verdcourt* 1158!
TANZANIA. Masai District: S. rim of Ngorongoro Crater, 12 June 1965, *Herlocker* 116! & Empakai Crater, top of E. rim, 17 Mar. 1973, *Frame* 99!; Rungwe District: Poroto Mts., Rungwe Massif, 14 Mar. 1914, *Stolz* 2565!
DISTR. U2, 3; K2–6; T2, 6, 7; Zaire, Ethiopia, Malawi, Mozambique and Zimbabwe
HAB. Upland grassland, *Acacia lahai* woodland and *Juniperus*, *Hagenia*, *Erica* forest and derived thicket, also clearings in bamboo forest; 1800–3200 m.
SYN. *C. bequaertii* De Wild. in Rev. Zool. Afr. 8, Suppl. Bot.: B18 (1920) & Pl. Bequaert. 2: 121 (1923). Type: Zaire, Ruwenzori, *Bequaert* 3590 (BR, holo.)
    C. longepetiolatum De Wild., Pl. Bequaert. 4: 15 (1926). Type: Zaire, Mukule, *Bequaert* 5904 (BR, holo.!)

var. **subalpinum** (*T.C.E. Fries*) *Verdc.*, comb. et stat. nov. Type: W. Mt. Kenya, 2900–3000 m., *R.E. & T.C.E. Fries* 1331 (S, holo. & iso.!)

Nutlets with external face glochidiate around margins and sparsely along a median line but with two smooth areas devoid of glochidia. Fig. 30/11, 12.

KENYA. Aberdare Range, near W. part of Nyeri track, 20 July 1948, *Hedberg* 1677!; N. Nyeri District: Mt. Kenya, Sirimon Track, 21 Sept. 1963, *Verdcourt* 3768!; Masai District: Narok, Nasampolai [Nasampulai] valley, 4 Sept. 1971, *Greenway & Kanuri* 14900!
TANZANIA. Moshi District: W. Kilimanjaro, Shira Plateau, Feb. 1928, *Haarer* 1144!; Morogoro District: Uluguru Mts., above Chenzema, 2 Jan. 1934, *Michelmore* 874! (intermediate); Mbeya District: Poroto Mts., road from Igoma to Kitulo, just below Kitulo [Elton] Plateau., 7 Feb. 1979, *Cribb et al.* 11327!
DISTR. K3, 4, 6; T2, 6, 7; Nigeria, Cameroon
HAB. Much as for typical variety; 2100–3300 m.
SYN. *C. lancifolium* Hook.f. in J.L.S. 7: 207 (1864); Bak. & Wright in F.T.A. 4(2): 53 (1905); Hutch. & Dalz., F.W.T.A. 2: 200 (1931). Types: Cameroon, Cameroon Mt., 2100–2400 m., *Mann* 1866 & 2004 (K, syn.!)
    C. amplifolium A. DC., forma *macrocarpum* Brand in E.P. IV. 252 (78): 141 (1921); Heine in F.W.T.A., ed. 2, 2: 324 (1963). Type as for last
    C. subalpinum T.C.E. Fries in N.B.G.B. 8: 415 (1923)
NOTE. Riedl considers that *C. subalpinum* is distinct, not because of the distribution of glochidia but the greater accrescence of the fruiting sepals but this seems variable also.

3. **C. lanceolatum** *Forssk.*, Fl. Aegypt.-Arab.: 41 (1775); C.B.Cl. in Hook.f., Fl. Brit. Ind. 4: 156 (1883), pro parte; Bak. & Wright in F.T.A. 4(2): 54 (1905), pro parte; Gürke & F. Vaupel in Z.A.E. 2: 280 (1911); De Wild. in B.J.B.B. 7: 38 (1920); Brand in E.P. IV. 252 (78): 37, fig. 18/A–G (1921), pro parte; F.P.N.A. 2: 130 (1947), pro parte; E.P.A.: 782 (1962); Heine in F.W.T.A., ed. 2, 2: 324 (1963); Taton in F.C.B., Borag.: 50 (1971), pro parte; Ivens, E. Afr. Weeds: 79 (1967); Kabuye & Agnew in U.K.W.F.: 521 (1974); Martins in F.Z. 7(4): 105, t. 30 (1990). Type: Yemen, Hadïe, *Forsskål* (C, holo.!)

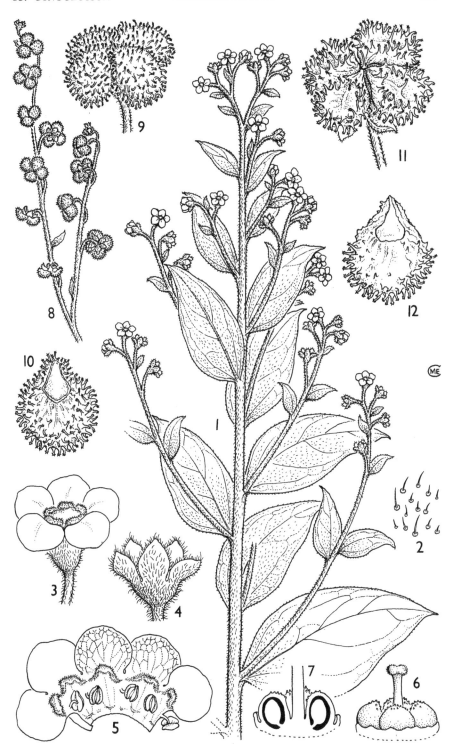

FIG. 30. *CYNOGLOSSUM AMPLIFOLIUM* var. *AMPLIFOLIUM*—**1**, habit, × ⅔; **2**, part of upper leaf surface, × 6; **3**, flower, × 4; **4**, calyx, × 6; **5**, corolla, opened out, × 6; **6**, ovary, × 20; **7**, same, longitudinal section, × 20; **8**, fruiting stem, × ⅔; **9**, fruit, × 3; **10**, nutlet, × 4. *C. AMPLIFOLIUM* var. *SUBALPINUM*—**11**, fruit, × 3; **12**, nutlet, × 4. 1–7, from *Battiscombe* 505; 8–10, from *Verdcourt* 1158; 11, 12, from *Verdcourt* 3768. Drawn by Mrs Maureen Church.

Annual or biennial (rarely? perennial) herb 0.2–1.2 m. tall, branched from the base or
± unbranched below but mostly with upper inflorescence-bearing part well-branched;
taproot thick, up to 20 × 2 cm.; stems drying brownish, with sparse to dense white
adpressed to spreading slightly tubercle-based hairs. Leaves elliptic, lanceolate or elliptic-
lanceolate, 2–13(–18) cm. long, 0.5–4(–5) cm. wide, acute at the apex, ± rounded at the
base, scabrid with tubercle-based hairs from rings of cystoliths, breaking off to leave
tubercles; venation ± impressed above, raised beneath; lower leaves with petioles
developed, upper ± sessile. Inflorescences terminal and axillary, usually characteristically
divaricately bifid, rapidly elongating, 1–20 cm. long, the individual branches mostly
(2.5–)4–13(–22) cm. long; flowers typically small; pedicels ± 1 mm. long up to 2.5 mm. and
mostly recurved in fruit. Sepals oblong, elliptic or ovate, 0.8–2 mm. long, 0.4–1.3 mm. wide,
± unequal and mostly 2 mm. long in fruit, with tubercle-based hairs outside. Corolla
usually white or sometimes pale blue or with a blue centre; tube 1 mm. long; lobes
rounded-ovate to oblong, 0.5–2 mm. long, 0.5–1 mm. wide; anthers 0.5 mm. long; style
0.2–0.3 mm. long. Fruits yellow-brown when ripe, 4–5 mm. wide, the nutlets subglobose-
obovoid, 1.5–2 mm. in diameter, evenly covered all over with slender glochidia over 0.5
mm. long, separating easily at maturity; fruiting style ± 0.5(–1.3) mm. long, narrowly
oblong and compressed, almost winged in dry state. Fig. 29/2, p. 105.

UGANDA. Karamoja District: Moroto, Jan. 1965, *J. Wilson* 1736!; Ankole District: km. 64 Masaka–
      Mbarara road, Mar. 1962, *Tallantire* 663!; Elgon, 22 May 1924, *Snowden* 886!
KENYA. Elgon, Oct. 1930, *Lugard* 175!; Kiambu District: Muguga, 28 Jan. 1963, *Verdcourt* 3577!; Teita
      Hills, E. of Wusi School, 8 Feb. 1966, *Gillett & B.L. Burtt* 17053!
TANZANIA. Ngara, 18 Dec. 1959, *Tanner* 4668!; Lushoto District: W. Usambaras, Soni, 25 Nov. 1972,
      Faulkner 4744!; Mpwapwa, 27 Feb. 1933, *Mr. & Mrs. Hornby* 480!
DISTR. U1–4; K3–7; T1–8; widespread from Arabia to S. & W. Africa and to India, Malaysia and
      China
HAB. Grassland, bushland, often as a weed in plantations, cultivation edges, murram roadsides and
      other areas of bare soil; (?600–)1140–2040(–?2520) m.

SYN. *C. canescens* Willd., Enum. Pl. Hort. Berol.: 180 (1809). Type, cult. hort. Paris (lecto.) (specimen in
      Willd. Herb. seems to be something different)
      *C. micranthum** Poir., Encycl. Méth. Bot., Suppl. 2: 431 (1811); A. DC. in DC., Prodr. 10: 149
          (1846). Type as for *C. canescens*
      C. hirsutum Thunb., Prodr. Fl. Cap.: 34 (1794); Willd., Sp. Pl., ed. 4, 1: 763 (1798); Jacq., Hort.
          Schoenbr. 4: 45, t. 489 (1804). Type: South Africa, 'Cape of Good Hope', *Thunberg* (UPS, holo.)
      Paracynoglossum lanceolatum (Forssk.) Mill in Notes Roy. Bot. Gard. Edin. 41: 474, figs. 1/c, k, 2,
          3B/a–b (1984)

NOTE. Typical specimens with very small white flowers and small subglobose nutlets coupled with
      rather characteristic branching are very easily recognised but there are many specimens which are
      intermediate with variants of *C. coeruleum* which may be hybrids or distinct infraspecific taxa (fig.
      29/3, p. 105). In Asia there are similar intermediates with other species. *C. lanceolatum* has been
      persistently misidentified and no attempt has been made to account for all the misidentifications
      even in the local literature. Mill transferred this species to *Paracynoglossum* partly on account of the
      nutlets being free from the style; Taton's key distinguishes it by the nutlets being coherent at the
      summit with the style but it appears from ripe material I have examined that this certainly is not
      always so. I have accepted some very small-flowered plants with blue flowers (e.g. *Geilinger* 3615,
      Ngorongoro, 11 Nov. 1932) as forms of *C. lanceolatum*. *Geilinger* 4959 (Kilimanjaro, N. Kibo, Rongai,
      25 Dec. 1932) is also a blue small-flowered form with few-flowered cymes not elongating into
      spike-like inflorescences; it definitely is not *C. coeruleum* subsp. *johnstonii* although *Geilinger* s.n.
      from the same locality is that species with clearly many-flowered scorpioid cymes and 1.8 mm. long
      styles.

4. **C. coeruleum** *A.DC. in DC.*, Prodr. 10: 148 (1846); Engl., Hochgebirgsfl. Trop. Afr.:
353 (1892); Bak. & Wright in F.T.A. 4(2): 53 (1905), pro parte; Brand in E.P. IV. 252(78): 146
(1921); E.P.A.: 781 (1962). Type: Ethiopia, Enschedcap, *Schimper* 542 (G, holo., BM, K, iso.!)

Perennial or less often annual or biennial herb (15–)25–120 cm. tall; stems rough with
± adpressed upwardly directed hairs above but spreading or deflexed beneath, very dense
on young parts; usually with a thick woody tap-root. Radical leaves narrowly ovate,
elliptic-oblong, linear-oblong or oblanceolate, 8–21(–30) cm. long, 1.1–6 cm. wide,
subacute at the apex, long-attenuate at the base, rough with rather sparse to dense short
adpressed white hairs from conspicuous white cystolith dots, less densely placed on the

*   Usually attributed to Desf., Tab. l'Ecole Bot. 220 (1804), but merely a nomen, the item in French
being merely the latin name translated.

undersurface; petiole 2–10 cm. long, spreading pubescent; cauline leaves similar, linear-lanceolate to ovate, 1.5–17(–30) cm. long, 0.2–5 cm. wide, subacute to acute at the apex, contracted to or slightly widened at the base, sessile or pseudopetiolate. Flowers in terminal and axillary, simple, dichasial or trifid cymes but not forming densely branched inflorescences; individual cymes short at first but soon 10–25 cm. long; pedicels mostly short, 2–7(–12) mm. long or that of central flower of dichasium up to 1.5 cm.; axes sparsely to densely adpressed pubescent. Calyx-lobes narrowly ovate to oblong or elliptic-obovate, 1.2–1.5(–3.5) mm. long, 0.4–1.2(–2) mm. wide, enlarging in fruit, 3–4 × 1.5–1.8 mm., acute to ± rounded at the apex, pubescent and ciliate and sometimes with marked cystolith-tubercles. Corolla blue or pinkish mauve turning bluish, sometimes with yellowish white centre or less often white or white with blue or pink centre; tube 1.5–2(–2.5) mm. long; lobes rounded-oblong, (1.2–)1.5–3 mm. long, 1–3 mm. wide. Style (0.3–)0.5–3 mm. long in fruit. Fruit often reddish, 6–9 mm. wide; nutlets compressed ovoid, (2.5–)3.5–4 mm. long, 3–3.5 mm. wide, covered all over with fairly short usually slender glochidia 0.5–1 mm. long and with few short additional tubercles above or with upper face almost or quite devoid of tubercles save for a median line, occasionally with peripheral ones joined at base to form a ± distinct margin but not as marked as in many species of the genus; there are, however, numerous intermediate fruit patterns.

SYN. *Paracynoglossum afrocaeruleum* Mill in Notes Roy. Bot. Gard. Edin. 41: 481, figs. 1/c, k, 2, 3, 13/a–b (1984), *nom. illegit.* (*C. coeruleum* A.DC. is not a later homonym of *C. coeruleum* Buch. Ham. ex D. Don, 1825, which appeared only in synonymy)

KEY TO INFRASPECIFIC VARIANTS

1. Nutlets with glochidia ± uniformly distributed on upper
   face 2
   Nutlets with glochidia ± restricted to margins and a median
   line . . . . subsp. **johnstonii** var.
   mannii
2. Leaves relatively larger and narrower, linear-lanceolate to
   lanceolate; **K** 3–5, **T** 1,2 . . . . subsp. **kenyense**
   Leaves broader, more broadly or shortly lanceolate to
   ovate or oblong-elliptic 3
3. Leaves oblong-elliptic, usually large, 1–18 × 0.5–5.5 cm.;
   style longer, ± 2.5 mm. long, and usually exserted from
   young fruiting calyx; **K** 1 . . . . subsp. **latifolium**
   Leaves elliptic-lanceolate, usually much narrower; style
   usually shorter, 0.5–2 mm. long, not exserted from
   young fruiting calyx; widespread . . . . subsp. **johnstonii** var.
   johnstonii

subsp **johnstonii** (*Bak.*) *Verdc.*, stat. nov. Type: Tanzania, Kilimanjaro, 1800 m., *Johnston* (K, holo.!, BM, iso.!)

Annual or perennial herb 0.3–1.2 m. tall; stems with adpressed apically directed hairs above and spreading hairs beneath. Radical leaves often not well developed particularly in annual plants but sometimes up to 19 × 4 cm., rarely to 30 × 5.5 cm., and long-petiolate; cauline leaves usually short and elliptic-lanceolate, less often linear-lanceolate, 2–22 cm. long, 0.8–5.5 cm. wide or lowest up to 30 × 5.5 cm., subacute to acute at the apex, contracted to sessile base, rough with upwardly directed hairs. Fruiting style often short, 0.5–2 mm. long. Nutlets with glochidia all over or limited to margins, a median line and underneath.

var. **johnstonii** (*Bak.*) *Bak. & Wright* in F.T.A. 4(2): 54 (1905). Type as for subsp. *johnstonii*

Nutlets with glochidia all over outer face. Less often annual. Fig. 29/4, p. 105.

UGANDA. Kigezi District: roadside W. of Muko, 1 Oct. 1961, *Rose* 1148!; Elgon, Kapchorwa, 8 Sept. 1954, *Lind* 430!; Mengo District: Entebbe, Oct. 1924, *Maitland* 271!
KENYA. Elgon, W. of Suam R., 24 Sept. 1966, *Bie* 66283!; N. Kavirondo District, by Yala R., 10 Dec. 1956, *Verdcourt* 1693!; Teita District: Vuria Hill, 8 May 1985, *Faden et al.* 118!
TANZANIA. Arusha, Sokon Native Court, 22 July 1956, *Milne-Redhead & Taylor* 11297!; Lushoto, 30 Mar. 1959, *Semsei* 2852!; Morogoro District: Uluguru Mts., Bunduki, 8 Sept. 1957, *Welch* 382!
DISTR. U2–4; K?2, 5–5, 7 (see note); T2, 3, 5–7; Cameroon, Zaire, S. Ethiopia, Malawi, Angola, distribution in Flora Zambesiaca area not fully worked out
HAB. Grassland with *Acacia*, overgrazed grassland and waste ground; 765–2400(–3650) m.

SYN. [*C. micranthum* sensu Oliv. in Trans. Linn. Soc., Bot., ser. 2, 2: 343 (1887), *non* Poir.]
   C. johnstonii Bak. in K.B. 1894: 29 (1894)
   [*C. lanceolatum* sensu Bak. & Wright in F.T.A. 4(2): 54 (1905), quoad *Volkens* 1661, *Buchwald* 227,
   Holst 3383, etc.; De Wild., Pl. Bequaert. 2: 122 (1923), pro parte; F.P.N.A. 2: 130 (1947), pro
   minore parte; Taton in F.C.B., Borag.: 50 (1971), pro minore parte, *non* Forssk.]
   [*C. coeruleum* sensu Martins in F.Z. 7(4): 104 (1990), pro parte *non* A.DC. sensu stricto]

NOTE. Kazmi (Journ. Arn. Arb. 52: 343 (1971)) included *C. johnstonii* in his synonymy of *C.
lanceolatum* but cannot I think have examined the type since it differs from *C. lanceolatum* in the
different inflorescences, usually blue flowers and larger fruits. Admittedly there are intermediates.
Many specimens with small flowers, particularly in eastern Uganda, appear to be hybrids with *C.
lanceolatum* but may be just a small-flowered taxon. *Faulkner* 4744 (Lushoto District, Soni, 25 Nov.
1972) with small fruits 5 mm. wide and divaricate inflorescences of blue flowers may also be a
hybrid. Occasional white-flowered specimens (? populations) of true var. *johnstonii* occur (e.g.
Archbold 371, Moshi District, Marangu, 18 Dec. 1963). In the Teita Hills populations of short annual
plants occur up to 50 cm. tall with ± sparse glochidia on the upper face of the nutlets and short
styles 0.8–1.1 mm. long. *Thorel* 3880 (Cherangani Hills, Embobut Forest, 3240 m.) a 20 cm. tall herb
with small narrow sessile leaves 3 × 0.7 cm. may belong here but is not in fruit. *Hooper & Townsend*
802 (Mbeya District, 15 km. from Mbeya on Chunya road, 16 Mar. 1975) has very small fruits with
short glochidia resembling those in true *C. coeruleum*. Whether it is a hybrid with *C. lanceolatum* or a
distinct variant is not certain.
   Undoubtedly the major problem is the relationship of some forms of var. *johnstonii* to Indian
species. B. L. Burtt (Notes Roy. Bot. Gard. Edin. 43: 350 (1986)) has independently noted this
problem but no one else appears to have drawn attention to it. *Mhoro* 522 (Iringa District, Kidatu, 8
Nov. 1971) is virtually indistinguishable from *C. wallichii* G. Don; *Greenway* 1661 (E. Usambaras,
Amani, 3 July 1929) and numerous specimens collected by *Peter* near Amani with leaves to 14 × 4.5
cm. and the very similar *Schlieben* 2221 (Mahenge, Sali, 18 May 1932) are also particularly close, e.g.
compare with *Sedgwick & Bell* 5420 from India. I have resisted pursuing this problem since it would
take a great deal of time and lead to much confusing change. Moreover, the large-flowered African
taxa are so different from those in India that it is doubtful any one would accept the logical
consequence. The nomenclature of the Indian species is very confused and the taxonomy far
from clarified (see Verdc. in K.B. 43: 343–348 (1988)). Future monographers, however, will not be
able to ignore the problem. *C. yemense* (Mill & A.G. Miller) Verdc.* is very close to some forms of *C.
johnstonii* and may be no more than another subspecies. A number of Peter sheets from the E.
Usambaras have narrower leaves; these sheets have been annotated var. A and had previously all
been referred to *C. lanceolatum* which they certainly are not, having bigger fruits, usually blue
flowers and different inflorescences although the flowers are small.

var. **mannii** (*Bak. & Wright*) Verdc., comb. nov. Type: Cameroon Mt., *Mann* 2005 (K, holo.! & iso.!)

Nutlets with glochidia mostly restricted to the margins and a median line across the outer face.
Often annual and calyx-lobes often more ovate and 'leafy'. Fig. 31/1, 2.

UGANDA. Kigezi District: Virunga Mts., Muhavura-Mgahinga saddle, 8 Nov. 1954, *Stauffer* 713!;
   Ruwenzori at 2100 m., *Doggett* !; Mbale District: Elgon, W. side, Jan. 1918, *Dummer* 3601! (var. with
   long fruiting style to 3 mm.)
KENYA. Northern Frontier Province: Ndoto Mts., Sirwan, 1 Jan. 1959, *Newbould* 3370!; ? District, 2nd
   day's march from Eldama [Eldoma] Ravine to Upper Mau Plateau, *Whyte* !; Meru District: Upper
   Imenti Forest, 28 June 1974, *R.B. & A.J. Faden* 74/916!
TANZANIA. Morogoro District: Mikese [Mikesse], 29 Apr. 1935, *Rounce* 384! (somewhat intermediate
   and said to have yellow flowers); Mbeya District; Poroto Mts., 15 May 1957, *Richards* 9713!; Rungwe
   Forest Reserve, June 1954, *Semsei* 1567!
DISTR. U2, 3; K1, 3, 4, 6; T2, 3 (intermediates), 4–7; Cameroon, Zaire, Rwanda and Burundi (fide
   Martins), Malawi, Mozambique, Zambia and Zimbabwe; also South Africa (Natal) (fide Martins)

HAB. Submontane forest, swamp forest, bamboo thicket and *Erica* associations, bushland,
   grassland, margins of cultivation and sometimes as a weed in resting fallows, etc.; 1740–3150 m.

SYN. *C. mannii* Bak. & Wright in F.T.A. 4 (2): 52 (1905); Hutch. & Dalz., F.W.T.A. 2: 200 (1931)
   C. geometricum Bak. & Wright in F.T.A. 4 (2): 52 (1905); De Wild., Pl. Bequaert. 2: 122 (1923);
   F.P.N.A. 2: 132 (1947); Martins in F.Z. 7(4): 107 (1990). Lectotype chosen by Hilliard & Burtt in
   Notes Roy. Bot. Gard. Edin. 43: 348 (1986): Malawi, Mt. Chiradzulu, *Whyte* (K, lecto.!)
   C. lanceolatum Forssk. subsp. *geometricum* (Bak. & Wright) Brand in E.P. IV. 252 (78): 140 (1921);
   Heine in F.W.T.A., ed. 2: 324 (1963); Taton in F.C.B., Borag.: 52, t. 7, map 18 (1971)
   C. lanceolatum Forssk. subsp. *geometricum* (Bak. & Wright) Brand var. *mannii* (Bak. & Wright)
   Brand in E.P. IV.252 (78): 140 (1921)
   [*C. lanceolatum* sensu De Wild., Pl. Bequaert. 2: 122 (1923), pro minore parte, *non* Forssk.]
   Paracynoglossum geometricum (Bak. & Wright) Mill in Notes Roy. Bot. Gard. Edin. 41: 478,
   fig. 1/d, j (1984)

---

* *Paracynoglossum yemense* Mill & A.G. Miller in Notes Roy. Bot. Gard. Edin. 41: 478 (1984)

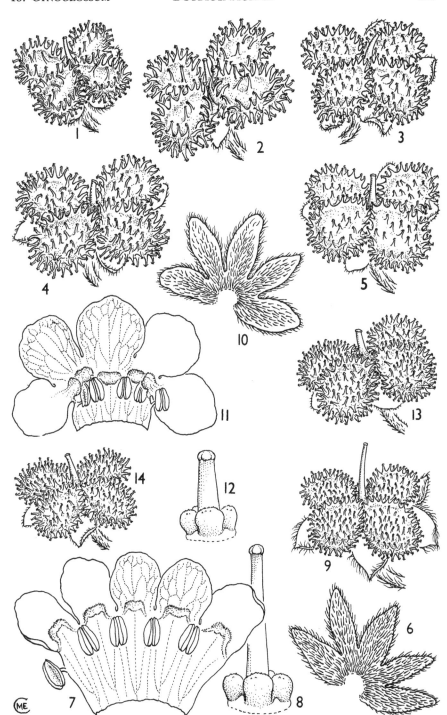

FIG. 31. *CYNOGLOSSUM COERULEUM* subsp. *JOHNSTONII* var. *MANNII*—**1**, **2**, fruits, × 5. *C. COERULEUM* subsp. *JOHNSTONII* vars. intermediate between var. *JOHNSTONII* and var. *MANNII*—**3–5**, fruits, × 5. *C. COERULEUM* subsp. *LATIFOLIUM*—**6**, calyx, × 8; **7**, corolla, × 8; **8**, ovary and style, × 20; **9**, fruit, × 5. *C. COERULEUM* subsp. *KENYENSE*—**10**, calyx, × 8; **11**, corolla, × 8; **12**, ovary and style, × 20; **13**, fruit, × 5; **14**, fruit, × 5. 1, from *Archbold* 2444; 2, from *Frame* 65; 3, from *Verdcourt* 1693; 4, from *Ngoundai* 131; 5, from *Jack* 19; 6–8, from *Bally* 5549; 9, from *Hepper & Jaeger* 6865; 10–12, from *Cooke* 45; 13, from *Symes* 241; 14, from *Verdcourt* 1498. Drawn by Mrs Maureen Church.

NOTE.   It has nearly always been suggested that plants with the nutlets bearing only a median line and margin of glochidia should be called *geometricum* either at specific rank or usually at subspecific rank. Brand treated them as a subspecies of *C. lanceolatum* but most workers at Kew have thought them much better placed as a variety of *C. coeruleum* which view is I think basically correct but undoubtedly once again obscured by possible hybridisation with *C. lanceolatum*. I have not employed the name *geometricum* for the variety since a varietal epithet is already available. Some plants determined as *C. geometricum* are not at all related, and here treated as distinct species (see species 6–8).

White-flowered forms occur, some of which at least are simple albinos, e.g. *Purseglove* 2929 (Kigezi District, Mt. Sabinio, June 1949) who specifically mentions them occurring in blue-flowered populations. Some with ± small white flowers, e.g. *Tanner* 5552 (Ngara District, Bushubi, Mu Rgwanza, 29 Dec. 1960) may be hybrids with *C. lanceolatum*. As with var. *johnstonii* some specimens of var. *mannii*, particularly from eastern Uganda, are very similar to some Indian taxa which display the same glochidial variation.

Numerous intermediates between var. *johnstonii* and var. *mannii* exist scattered throughout the Flora area and make varietal rank the only way of dealing with the problem; moreover non-fruiting specimens cannot be named to variety, e.g. *Tweedie* 4104 (Kenya, Kitale, E. of Milimani, Aug. 1971, 1800 m.). Fig. 31/3–5, p. 111.

subsp. **latifolium** *Verdc.*, subsp. nov., a subsp. *johnstonii* foliis caulinis oblongo-ellipticis plerumque majoribus 1–15 cm. longis 0.5–5 cm. latis, stylo post anthesin 2.5–3 mm. longo, basi dilatato differt. Typus: Kenya, Northern Frontier Province, Mt. Kulal, *Bally* 5549 (K, holo.!, EA, iso.!)

Perennial herb 0.4–1 m. tall, rough with spreading hairs. Radical leaves oblanceolate to broadly elliptic, 9 cm. long, 2.5–6 cm. wide; petiole ± 8 cm. long; cauline leaves elliptic-oblong, mostly ± large, 1–18 cm. long, 0.5–5.5 cm. wide, acute, attenuated at the base, very rough above with dense white cystolith spots and rather short white hairs, with denser longer hairs beneath. Corolla 7–8 mm. wide. Fruiting style 2.5–3 mm. long, widened beneath. Fruit with slender glochidia all over. Fig. 31/6–9, p. 111.

KENYA.   Northern Frontier Province: Marsabit, 14 Feb. 1953, *Gillett* 15102! & near Marsabit town, 14 May 1970, *Magogo* 1324! & Mt. Kulal, Gatab, 17 Nov. 1978, *Hepper & Jaeger* 6865!
DISTR.   **K**1; ? Ethiopia
HAB.   Evergreen forest, derived thickets of former *Juniperus* forest, also rocky places in grassland; 1200–2100 m.

SYN.   [*C. coeruleum* sensu Hepper et al., Annot. Check-List Pl. Mt. Kulal: 86 (1981), *non* A.DC. sensu stricto]
[*C. amplifolium* sensu Hepper et al., Annot. Check-List Pl. Mt. Kulal: 86 (1981), *non* A.DC.]

subsp. **kenyense** *Verdc.*, subsp nov. a subsp. *johnstonii* foliis pro rata longioribus et angustioribus, floribus plerumque majoribus; a subsp. *coeruleo* A.DC. glochidiis nuculae longioribus gracilioribus 0.5–1 mm. longis, foliis pro rata longioribus, manifeste 3-nervatis differt. Typus: Kenya, Uasin Gishu District, Eldoret, *L.A. Cooke* 45 (K, holo.!)

Perennial herb (15–)30–60(–100) cm. tall, hispid with upwardly directed hairs and usually spreading hairs at base. Radical leaves oblanceolate, 8–15 cm. long, 1–1.8 cm. wide; petiole 2–7 cm. long; cauline leaves linear-lanceolate to lanceolate, 2–15(–17) cm. long, 0.2–1.5(–2.5) cm. wide, subacute to acute at apex, contracted to sessile base, rough with upwardly directed hairs from conspicuous cystolith spots. Fruiting style 0.8–1.5(–1.8) mm. long, ± flattened. Fruit entirely covered with glochidia. Fig. 31/10–14, p. 111.

KENYA.   Trans-Nzoia District: Elgon, 13 Nov. 1957, *Symes* 241!; Lake Naivasha, 10 June 1956, *Verdcourt* 1498!; Kiambu District: Kabete, 28 May 1947, *Bogdan* 560!
TANZANIA.   Musoma District: Limuta Hill, 3 Mar. 1964, *Myles Turner* in *E.A.H.* 12964!; Masai District: Ngorongoro, 21 Nov. 1956, *Tanner* 3273!; Mbulu District: Oldeani Mt., 19 Nov. 1957, *Tanner* 3813!
DISTR.   **K** 3–6; **T** 1, 2; not known elsewhere (see note)
HAB.   Grassland, grassland with scattered trees, forest edges, also in plantations and sometimes as a weed; 1440–3150 m.

SYN.   [*C. coeruleum* sensu Bak. & Wright in F.T.A. 4 (2): 53 (1905), pro parte quoad *Kassner* 752, etc.; Brand in E.P. IV.252 (78): 146 (1921), pro parte; Kabuye & Agnew in U.K.W.F.: 523 (1974), *non* Hochst. ex A.DC. (1846) sensu stricto]
[*C. lanceolatum* sensu Brand in E.P. IV.252 (78): 139 (1921), quoad *Scheffler* 246, etc., *non* Forssk.]
C. coeruleum A.DC. var. *winkleri* Brand in E.P. IV.252 (78): 146 (1921). Type: between Nairobi ["Mairoti"] and the Rift, *Winkler* 4183 (?B, holo.†, WRSL, iso.!)
C. sp. A sensu Kabuye & Agnew in U.K.W.F.: 521 (1974)

NOTE.   There is great variation in flower size, e.g. *Greenway* 13545 (cult. Hort. Greenway, Kiambu District, Muguga, 21 Dec. 1968) has the corolla-limb 7–8 mm. wide. Material cited under sp. A (e.g. *Hedberg* 19 from Elgon) by Kabuye & Agnew is I think only a local white-flowered variant; such variants seem common in western Kenya. *Tweedie* 3737 (W. Suk District, Kapenguria, Dec. 1969) is a small white-flowered plant of wet savannah but is not in fruit. *C. yemenense* (Mill & A.G. Miller) Verdc. is very close to subsp. *kenyense*. Populations from the Mbulu Highlands are better placed in subsp. *kenyense* than in subsp. *johnstonii* but the situation is obscured here by intermediates.

5. **C. aequinoctiale** *T.C.E. Fries* in N.B.G.B. 8: 416, fig. 6 (1923). Type: Kenya, E. Mt. Kenya, *R.E. & T.C.E. Fries* 303 (S, holo.!, K, photo.)

Perennial herb 30–80 cm. tall with slender rootstock; stems pubescent with ± short rather irregularly positioned white hairs. Radical leaves petiolate, narrowly-elliptic to elliptic-lanceolate, 5–15 cm. long, 1–2 cm. wide, acute at the apex, attenuate at the base, pubescent with adpressed ± short ± downwardly directed white hairs mostly from cystolith spots; petioles 2–7 cm. long; stem-leaves rather sparse, mostly linear-lanceolate, 2.5–5.5 cm. long, 0.4–1 cm. wide, narrowly acute at apex, sessile. Flowers in axillary and terminal ± corymb-like ± 10-flowered cymes; peduncles 1–8 cm. long; pedicels 0.2–1.7 cm. long; all axes densely hispid with upwardly directed pale fulvous hairs. Calyx-tube obconic; lobes broadly ovate, 1.5 mm. long, 1.2–1.8 mm. wide, rounded or occasionally subacute in fruit, covered with upwardly directed white and pale fulvous rather coarse hairs, often conspicuously tuberculate at the base, ciliate. Corolla blue, drying dark brown; tube 1.5–2 mm. long; lobes ± 2 mm. long and wide. Style short, 0.4–1 mm. long. Fruit 0.9–1 cm. wide; nutlets compressed-ovoid, 4–5 mm. long, completely uniformly covered with glochidia under 0.5 mm. long. Fig. 32/1, p. 114.

UGANDA. Mbale District: Mt Elgon, Butandiga to Bulambuli, Apr. 1929, *Saundy & Hancock* 95!
KENYA. ?Elgeyo District: Cherangani Hills, Kapseis, 6 Aug. 1968, *Thulin & Tidigs* 77!; Trans-Nzoia District: Elgon, *Irwin* 12!; Nakuru District: Nyahururu [Thomson's] Falls, 12 Aug. 1950, *Lacey* 24A!
DISTR. **U** 3; **K** 2–4; not known elsewhere (see note)
HAB. Upland grassland over basement complex; 2100–2870 m.

SYN. [*C. coeruleum* sensu Bak. & Wright in F.T.A. 4(2): 53 (1905), quoad *Johnston* s.n., *non* A.DC.]
C. *sp. B* sensu Kabuye & Agnew in U.K.W.F.: 521 (1974) e descr.

NOTE. Johnston describes the flowers as "chrome green". *Myers* 11141! (Sudan, Equatoria, Didinga Mts., 1950 m., 26 Apr. 1939) is similar to *C. aequinoctiale* but the sepals differ somewhat and no fruits are available. S. Martins (F.Z. 7(4): 104 (1990)) records this species from Zambia (Mbala) but this seems a strange distribution. I have examined the material and, although similar, doubt if it is conspecific. Unfortunately it is not in fruit; further collections will solve the problem.

6. **C. hanangense** *Verdc.*, sp. nov., affinis *C. coeruleo* A.DC. var. *johnstonii* (Bak.) Bak. & Wright sed corolla majore, tubo 2.5–3 mm. longo, lobis 3 × 3 mm. stylo 1.4–2.5 mm. longo differt. Type: Tanzania, Mbulu District, Mt. Hanang [Guruwe], *B.D. Burtt* 2264 (K, holo.!, EA, iso.)

Biennial herb (*fide* Greenway) to 90 cm.; stems with upwardly directed adpressed hairs, dense near the inflorescences, coarsely spreading hairy beneath. Radical leaves oblanceolate, up to 15 cm. long, 1.8 cm. wide, tapering acute at the apex, very gradually attenuated into 7 cm. long petiole; cauline leaves lanceolate, 2.5–11 cm. long, 0.3–1.5 cm. wide; all with white ± adpressed hairs from cystolith spots. Inflorescences terminal and axillary and on lateral leafy branches, at first condensed but later lax; peduncles up to 10 cm. long; pedicels 2 mm. long. Calyx-lobes lanceolate, finally in fruit up to 4–5 × 1.5 mm. Corolla bright blue; tube 2.5–3 mm. long; limb 7.5 mm. wide, with lobes 3 × 3 mm.; throat-bosses prominent and squarish, ± 1 mm. long. Style 1.4–2.5 mm. long. Ripe fruits not seen but nutlets at least 3 mm. wide with glochidia around the margin and across a median line, with very few on intervening faces. Fig. 32/2, p. 114.

TANZANIA. Mbulu District: Mt. Hanang, 26 Dec. 1929, *B.D. Burtt* 2264! & same, NE. slopes, 8 Feb. 1946, *Greenway* 7675! & same, July 1952, *Brooks* 100!
DISTR. **T**2; not known elsewhere
HAB. Upland grassland; 2340–3000 m.

NOTE. On existing evidence I have preferred to treat this as a species rather than suggest relationships with large-flowered variants of *C. coeruleum* subsp. *kenyense* which occur on several Kenya mountains.

7. **C. ukaguruense** *Verdc.*, sp. nov., affinis *C. coeruleo* A.DC. sed sepalis demum lanceolatis 4–5 mm. longis, 0.9–1.7 mm. latis et simul pedicellis longioribus usque 0.8–3.2 cm. longis differt. Type: Tanzania, Kilosa District, Ukaguru Mts., *Mabberley* 1341 (K, holo.!, DSM, iso.!)

Herb; rootstock and lower leaves unknown; young stems hispid with adpressed upwardly directed hairs, older ones slender, distinctly woody, glabrous, with a dark brown epidermis. Stem-leaves narrowly elliptic, 2–9 cm. long, 0.5–2.2 cm. wide, acute, cuneate or

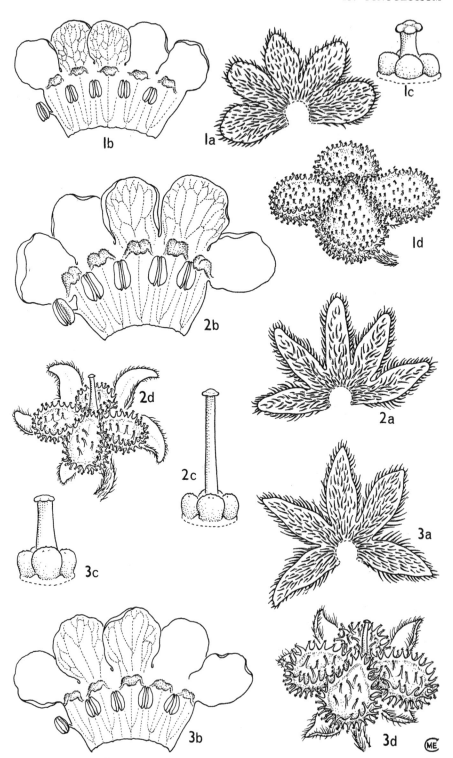

FIG. 32. *CYNOGLOSSUM SPP.*—**a**, calyx, × 8; **b**, corolla, × 8; **c**, ovary and style, × 20; **d**, fruit, × 5 of **1**, *C. AEQUINOCTIALE*; **2**, *C. HANANGENSE*; **3**, *C. UKAGURUENSE*. 1a–c, from *Ravell* 219; 1d, from *Lacey* 24a; 2a–c, from *Burtt* 2264; 2d, from *Greenway* 7675; 3, from *Mabberley* 1341. Drawn by Mrs Maureen Church.

narrowly attenuate at the base, scabrid above with short adpressed white hairs from cystolith spots and similar hairs but no spots beneath; apparent petiole 0–1 cm. long. Cymes few-flowered, terminal and on upper axillary shoots, 4–15 cm. long, usually one flower well below the rest close to upper node; pedicels at first ± 5 mm. long, lengthening to 3.2 cm., longest in the lowermost flower. Sepals ± oblong-lanceolate at first, ± 3 mm. long, becoming lanceolate, 4–5 mm. long, 0.9–1.7 mm. wide, ± narrowly rounded or acute, hispid outside, the hairs at base longer and from cystolith spots. Corolla blue; tube 2 mm. long; lobes ± round, 2.5 mm. long and wide; bosses ± 1 mm. wide; anthers 0.7 mm. long; style 1 mm. long lengthening to 2 mm. in fruit. Fruits ± 9 mm. wide, the nutlets rounded-ovate, contracted to the base, ± 4 mm. long and wide, densely glochidiate on base and sides but much sparser on the faces although not confined to a median line; scar of attachment ± obtriangular, 1.5 mm. long, continuing at base into a portion of the split style. Fig. 32/3.

TANZANIA. Kilosa District: Ukaguru Mts., Mamiwa Forest Reserve, ridge to N. of Mandege Forest Station, 2 Aug. 1972, *Mabberley* 1341!
DISTR. **T**6; not known elsewhere
HAB. Secondary forest with *Albizia, Myrianthus, Maesa, Dombeya*, etc.; 1500–1650 m.

8. **C. cheranganiense** *Verdc.*, sp. nov., affinis *C. coeruleo* A.DC. var. *mannii* (Bak. & Wright) Verdc., sed habitu elatiore, foliis longioribus, inflorescentiis ramosioribus, glochidibus nuculae brevibus triangularibus, stylo 1.8–2 mm. vel ultra longo differt. Typus: Kenya, Elgeyo District, Cherangani Hills, Kamalagon [Kameligon], *Mabberley & McCall* 207 (K, holo.!)

Perennial (or possibly sometimes annual) herb to 1.8 m. tall from a thick tap-root; stems strict and sparsely branched, often streaked with purple above. Leaves narrow and elongate; radical leaves oblanceolate, up to ± 30 cm. long, 3.5 cm. wide, narrowly acute at the apex, strongly attenuated at the base so as to appear long-petiolate; cauline leaves lanceolate to oblong-lanceolate, 4–20 cm. long, 0.7–2 cm. wide, the uppermost sessile, the lower gradually more attenuate until similar to radicals; both surfaces sparsely to fairly densely covered with white hairs arising from cystolith spots. Flowers in long-pedunculate well-branched compound dichasia of scorpioid cymes which in the early stages are ± capitate and distinctive; cymes ultimately ± 12 cm. long; peduncles up to 16 cm. long; pedicels 1.5–3 mm. long; all axes densely white pubescent. Calyx-lobes oblong-elliptic, 2–3 mm. long, ± 1 mm. wide, distinctly ciliolate, pubescent outside with hairs from tubercular cystolith bases, very distinct near base. Corolla deep blue, white at throat, rarely all white; tube 2.5–3.5 mm. long; lobes ± round, 2–4 mm. long and wide; style (1.3–)1.8 mm. long to over 2 mm. long in fruit, slender. Nutlets greyish, ovate in plan, 2–2.2 mm. long, 1.8–2 mm. wide, with short triangular glochidia on the back and around the margin and 2 or 3 across the median ridge, also a few shorter ones between. Fig. 33/1–4, p. 116.

KENYA. Elgeyo District: Cherangani Hills, Kaisungur [Kaisungor], Feb. 1965, *Tweedie* 2984! and 1°5′N, 35°26′E, 21 Sept. 1949, *Maas Geesteranus* 6319! & without precise locality, Sept. 1934, *Dale* in *F.D.* 3270!
DISTR. **K** 2 (see note), 3, 6 (see note); not known elsewhere
HAB. *Erica arborea, Cliffortia, Protea* scrub; grassland with scattered bamboo, *Juniperus* and *Hagenia* clumps; 2850–3270 m.
NOTE. *Agnew et al.* 10508 (W. Suk District, Sekerr Mt., 4 Aug. 1968 at 2640 m.), *Greenway & Kanuri* 14529 (Masai District, Nasampolai [Enesambulai] valley, 25 July 1970 at 2400 m.) and *Glover et al.* 1469 (Masai District, about 11.2 km. from Cobb's Gate, Toboti, near edge of Mau Forest Reserve at 2850 m.) seem to be the same species but further study is needed.

9. **C. karamojense** *Verdc.*, sp. nov., affinis *C. cheranganiensis* Verdc. sed caulibus foliisque pilis albidis ± longis ± dense obtectis, nuculis 1–4 sursum inclinatis haud profunde crateriformibus ambitu rotundatis vel triangularibus margine ± uniseriatim glochidiatis medio sparse glochidiatis inferne minute tuberculatis alibi laevibus, glochidibus marginatis anguste triangularibus differt. Typus: Uganda, Karamoja District, Mt. Morongole, *Dawkins* 807 (K, holo.!, ENT, EA, iso.)

Perennial herb to 60 cm., presumably from a woody rootstock; stems hirsute with dense ± adpressed downwardly directed and some ± spreading white hairs. Leaves elliptic-lanceolate to lanceolate; radical up to 14 cm. long, 2.5–5 cm. wide, narrowly acute at the apex, attenuate at base into a petiole ± 8 cm. long; cauline 2.5–12.5 cm. long, 0.3–2.5 cm.

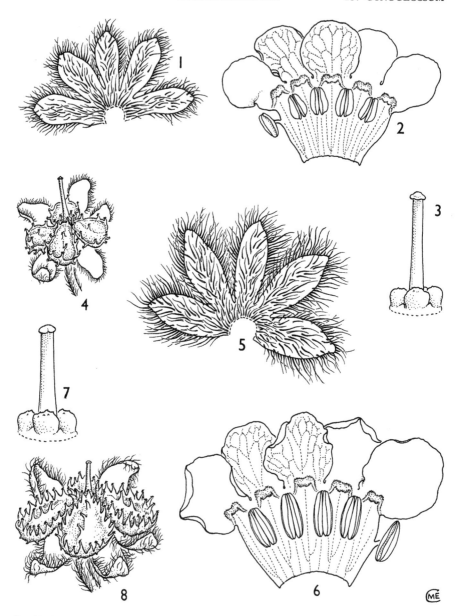

FIG. 33. *CYNOGLOSSUM CHERANGANIENSE*—**1**, calyx, × 8; **2**, corolla, × 8; **3**, ovary and style, × 20; **4**, fruit, × 5. *C. KARAMOJENSE*—**5**, calyx, × 8; **6**, corolla, × 8; **7**, ovary and style, × 20; **8**, fruit, × 5. 1–3, from *Dale* 3270; 4, from *Tweedie* 2984; 5–8, from *Champion* s.n. Drawn by Mrs Maureen Church.

wide, narrowly acute at apex, the upper quite sessile and ± rounded at base, the lower attenuate at base into an apparent petiole which widens at base and is often ± amplexicaul, all with ± dense ± long white hairs, similar to stem, particularly on the venation beneath. Flowers in well-branched inflorescences of terminal and axillary dichasial or trifid cymes, the individual branches up to 7 cm. long, all very condensed in young state; axes densely pubescent; pedicels 1–2 mm. long. Sepals narrowly-oblong or elliptic, 2.5–3 mm. long, 0.8–1.8 mm. wide, densely pubescent, tending to be connivent when corolla has dropped, pubescent outside and long-ciliate. Corolla bright blue; tube ±

2.5 mm. long; lobes rounded, 2.5–3 mm. long, 2.5 mm. wide. Style 1–2 mm. long. Fruits ± 8 mm. wide; nutlets 1–4, inclined upwards towards style, shallowly bowl-shaped, round to triangular in plan, 4–5 mm. long, 3–4 mm. wide, with a ± uniseriate row of marginal flat narrowly-triangular glochidia and a few on a median line and with small tubercles on the under surface. Fig. 33/5–8.

UGANDA. Karamoja District: Mt. Moroto, Jan. 1959, *J. Wilson* 649! & Mt. Morongole, 11 Nov. 1939, *A.S. Thomas* 3293! & June 1946, *Eggeling* 5650! & 29 June 1953, *Dawkins* 807!
KENYA. Turkana District: Mt. Kachagalau [Kachonkulu], Feb. 1933, *Champion* !
DISTR. U 1; K 2; not known elsewhere
HAB. Upland grassland with *Protea*, etc., sometimes in rock crevices; 2400–2940 m.

16. AFROTYSONIA*

Rauschert in Taxon 31: 558 (1982); Mill in Notes Roy. Bot. Gard. Edin. 43: 467–75 (1986)

Tysonia H. Bolus in Hook., Ic. Pl. 20, t. 1942 (1890) *non* Fontaine (1889)

Perennial herbs with single unbranched stems clothed with petiole remnants; rootstocks horizontal or subvertical. Basal leaves broadly ovate to elliptic-lanceolate, long-petiolate; cauline leaves elliptic to narrowly lanceolate the lower long-petiolate, the upper sessile. Flowers in large ± lax branched panicles with reduced bracts or lower leaf-like, the individual cymes 8–15-flowered, simple or branched, pedunculate; pedicels elongating considerably in fruit. Calyx divided nearly to the base into 5 equal lanceolate persistent but scarcely accrescent lobes. Corolla white, sometimes tinged mauve or yellowish, campanulate or subrotate; lobes ovate, obtuse. Throat-bosses exserted, trapeziform, oblong-triangular or semilunar, usually emarginate. Filaments inserted in middle of corolla-tube, ± exserted; anthers almost medifixed, versatile. Nectary scales ± 10 at base of corolla-tube, each with 2 divergent horns. Ovary indistinctly 4-lobed; style filiform, ± equalling the stamens; stigma small, capitate. Nutlets 1–4, all equal or one much exceeding the others, either wingless and densely glochidiate or with a broad cartilaginous ± undulate wing but lacking glochids.

Three species, two in South Africa, the other widely disjunct occurring in Tanzania. Recognised at Kew as a member of the genus nearly 30 years ago, it has only very recently been described.

A. pilosicaulis Mill in Notes Roy. Bot. Gard. Edin. 43: 472 (1986). Type: Tanzania, Ufipa District, Sumbawanga, Mbisi Forest, *Richards* 8679 (K, holo.!)

Herb at least 35 cm. tall, the stems densely leafy and densely shortly adpressed pilose with retrorse rather bristly hairs. Basal leaves elliptic-lanceolate to ovate-lanceolate, 9–12 cm. long, 3 cm. wide, eventually becoming 19 × 5 cm., attenuated into a petiole about equalling the lamina, ± densely adpressed hispid-setulose; cauline leaves sessile, elliptic or ovate-lanceolate, 12–13 cm. long, 4 cm. wide. Inflorescences densely leafy; bracts leaf-like, the lower 8–11 cm. long, 2.5–4 cm. wide; secondary bracts linear-lanceolate. Pedicels 6–9 mm. long, densely adpressed pilose. Calyx-lobes 3–5 mm. long, acute, grey-pilose. Corolla white, shortly campanulate; tube 5.5–6 mm. long, the bosses white, exserted, (0.6–)0.9–1.5 mm. long, 2–2.2(–2.5) mm. wide; lobes broadly ovate-oblong, 2.5–3.5 mm. long, ± spreading, emarginate. Filaments 2.5–3 mm. long, scarcely exserted. Style (3–)4–6(–7) mm. long. Nutlets not known. Fig. 34, p. 118.

TANZANIA. Ufipa District: Sumbawanga, Mbisi Forest, 13 Mar. 1957, *Richards* 8679! & Malonje, 19 July 1962 (leaves only), *Richards* 16809!
DISTR. T 4; not known elsewhere
HAB. Rough grassland; 2100–2400 m.

NOTE. Further material including fruits is much needed.

---

* Based entirely on Mill's account

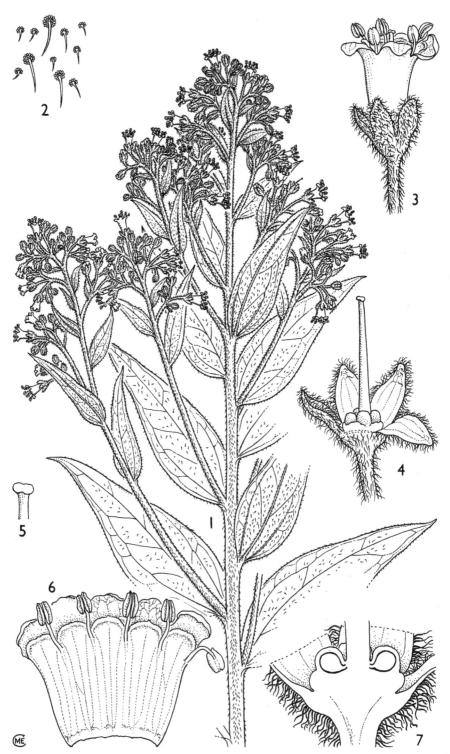

FIG. 34. *AFROTYSONIA PILOSICAULIS*—**1**, habit, × ⅖; **2**, part of lower leaf surface, × 10; **3**, flower, × 4; **4**, flower with corolla and one calyx-lobe removed, × 5; **5**, stigma, × 12; **6**, corolla, opened out, × 5; **7**, ovary, longitudinal section, × 12. All from *Richards* 8679. Drawn by Mrs Maureen Church.

Addendum

15a. **Heliotropium applanatum** *Thulin et Verdc.*, sp. nov. *H. strigosi* Willd. valde affinis sed habitu perfecte prostrato, inflorescentiis valde reductis differt. Typus: Somalia, Bay, Buur Diinsor, about 3 km. SW. of Diinsoor, *Thulin et al.* 7605 (UPS, holo.!, K, MOG, iso.!)

Completely prostrate delicate much-branched annual, 6–25 cm. wide; stems and foliage strigose with white hairs. Leaves elliptic to narrowly elliptic-oblong, 0.5–1.7 cm. long, 2–3 mm. wide, acute. Inflorescences 0.5–1.5 cm. long, mostly short. Sepals lanceolate, 2–3.5 mm. long, 0.5–1.5 mm. wide. Corolla white, 2.5–3 mm. long, the tube ± 1 mm. long and limb ± 2.5–3 mm. wide. Style and stigma as in *H. strigosum*. Nutlets ovoid-segment-shaped, orange-brown turning black, 1.2 mm. long, 1 mm. wide, with pits on the contiguous flat faces, with very short hairs on curved outer face.

KENYA. Northern Frontier Province: 36 km. N. of Wajir on Tarbaj road, 14 May 1974, *Gillett & Gachathi* 20672! & Wajir, 22 June 1951, *Kirrika* 67! & 8 km. E. of Wajir, on road to Wajir Bor, 1 June 1977, *Gillett* 21296!
DISTR. **K**1; Somalia
HAB. *Acacia, Commiphora, Delonix* bushland or open woodland on red sand; 220–420 m.

NOTE. Thulin has seen this growing together with *H. strigosum* and confirms what I suspected that the two are specifically distinct. *Kirrika* 67 is said to be from 2000' but Wajir is only 240 m. *Gilbert & Thulin* 1654, Mandera, 12 km. S. of El Wak on Wajir road, 11 May 1978, belongs here although formerly it had been suggested it might be a hybrid between *H. strigosum* and *H. sessilistigma*.

New names validated in this fascicle

Cordia crenata Del.
  subsp. **shinyangensis** *Verdc.*, 20
Cordia guineensis *Thonn.*
  subsp. **mutica** *Verdc.*, 25
Cordia peteri *Verdc.*, 10
Cordia trichocladophylla *Verdc.*, 28
Cynoglossum amplifolium *A.DC.*
  var. **subalpinum** (*T.C.E. Fries*) *Verdc.*, 106
Cynoglossum cheranganiense *Verdc.*, 115
Cynoglossum coeruleum *A.DC.*
  subsp. **johnstonii** (*Bak.*) *Verdc.*, 109
    var. **mannii** (*Bak. & Wright*) *Verdc.*, 110
  subsp. **kenyense** *Verdc.*, 112
  subsp. **latifolium** *Verdc.*, 112
Cynoglossum hanangense *Verdc.*, 113
Cynoglossum karamojense *Verdc.*, 115
Cynoglossum ukaguruense *Verdc.*, 113
Ehretia glandulosissima *Verdc.*, 36
Ehretia janjalle *Verdc.*, 36
Heliotropium *L.*
  sect. **Gottliebia** *Verdc.*, 50
  sect. **Rutidotheca** (*A.DC.*) *Verdc.*, 50
Heliotropium applanatum *Thulin & Verdc.*, 119
Heliotropium bullockii *Verdc.*, 75
Heliotropium longiflorum (*A.DC.*) *Jaub. & Spach*
  subsp. **undulatifolium** (*Turrill*) *Verdc.*, 68
Heliotropium pectinatum *F. Vaupel*
  subsp. **harareense** (*Martins*) *Verdc.*, 59
  subsp. **mkomaziense** *Verdc.*, 61
  subsp. **septentrionale** *Verdc.*, 61
Heliotropium rariflorum *Stocks*
  subsp. **hereroense** (*Schinz*) *Verdc.*, 70
Heliotropium steudneri *Vatke*
  subsp. **bullatum** *Verdc.*, 65
  subsp. **steudneri**
    var. **iringensis** *Verdc.*, 65
Trichodesma *R.Br.*
  sect. **Serraticaryum** *Verdc.*, 92

GEOGRAPHICAL DIVISIONS OF THE FLORA

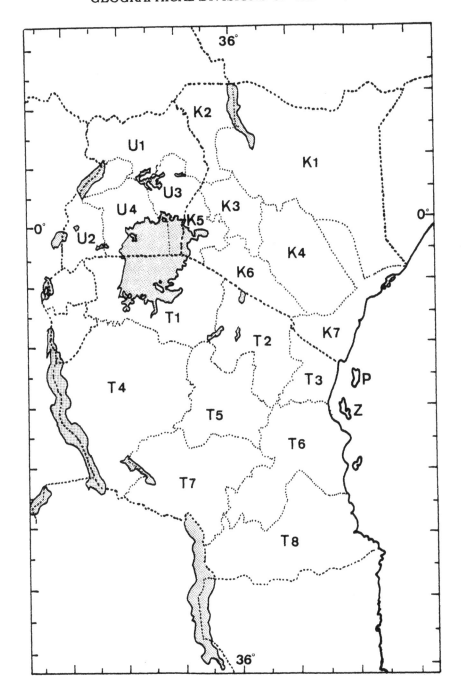

Milton Keynes UK
Ingram Content Group UK Ltd.
UKHW031133141024
449569UK00006B/208